THE FIRES OF CULTURE

Energy Yesterday and Tomorrow

DATE DUE			
Apr13 '80			

THE FIRES OF CULTURE

Energy Yesterday and Tomorrow

Carol Steinhart

John Steinhart

Duxbury Press

North Scituate, Massachusetts

A division of Wadsworth Publishing Co.

Belmont, California

Duxbury Press

North Scituate, Massachusetts

A DIVISION OF WADSWORTH PUBLISHING COMPANY, INC.

The Fires of Culture: Energy Yesterday and Tomorrow was edited and prepared for composition by Maryellyn Montoro. The interior design was provided by Jane Lovinger and the cover was designed by Designworks, Inc.

L.C. Cat. Card No.: 74-84840
ISBN 0-87872-079-0

PRINTED IN THE UNITED STATES OF AMERICA

1 2 3 4 5 6 7 8 9 10 — 79 78 77 76 75 74

To: Martha, Geoff, and Gail Steinhart,
who will have to live with the outcome.

CONTENTS

PREFACE

When we began this book in 1971, national concern over environmental deterioration had resulted in the passage of legislation to mitigate losses in air and water quality. Yet it seemed to us that many or most environmental problems were associated with the increasing energy use in society — and there was no direct policy for energy supplies or use. It had become obvious how precariously all industrial societies were balanced on the assumption of endlessly expanding supplies of cheap energy. Events since then have provided harsh proof of the consequences of this assumption. Availability and prices of energy have changed our lives, have realigned old loyalties, and have played the key role in foreign policy. The vulnerability of industrial countries is now clear for all to see.

As we write this preface, the Arab oil embargo has been lifted and gasoline flows freely once more although its price is high. Former President Nixon relegated the energy crisis to the status of an energy problem. In the minds of many, even the problem has been solved. But shortages of fuel and other forms of energy will be with us for some time. In order to choose among the options available, a wide range of consequences needs to be examined. To provide broad, nontechnical coverage in a book of modest size, we have treated many complex subjects in a brief and simplified manner. The issue of exponential growth and its implications, for example, is intimately entwined with future energy supply and use, but we have dealt with it only briefly. Many choices that must be made depend far more on hopes and dreams than on objective "facts." For this reason, we have tried to be fair when presenting the evidence, but to make our philosophical biases clear. There appears to be no way out of our present difficulties that is cheap, safe, and delightful.

Many people have contributed to the ideas and material presented in this book, through stimulating and sometimes heated discussions, through providing us with the results of their own work, and through thoughtful criticism of the manuscript at various stages in its development. To all of these, our sincere thanks. We have profited from their suggestions and our book has been improved by their efforts. But we have not always taken their advice, and we accept responsibility for whatever defects remain.

Carol E. Steinhart
John S. Steinhat

Madison, Wisconsin
August 1974

UNITS USED IN THIS VOLUME

A.   UNITS OF ENERGY

    Calorie. The amount of heat required to raise the temperature of 1 gram of 1° Centigrade (or 0.035 ounces of water 1.8° Fahrenheit).

    Kilocalorie. One thousand calories. The familiar food "calorie" is actually a kilocalorie.

    Kilowatt-hour. The amount of energy expended or work performed in one hour at a rate of one kilowatt, equivalent to 860 kilocalories. A kilowatt-hour of electricity will operate a 100 watt light bulb for ten hours.

B.   UNITS OF POWER (the rate at which energy flows or work is done; power is energy per unit time).

    Watt. 0.239 calories per second.

    Kilowatt. One thousand watts; 0.239 kilocalories per second.

    Megawatt. One million watts; one thousand kilowatts; 239 kilocalories per second.

C.   UNITS OF ELECTROMOTIVE FORCE or potential difference.

    Volt. The electric potential difference across a conductor which carries a current of 1 ampere and dissipates heat energy at the rate of 1 watt.

    Kilovolt. One thousand volts.

D.   UNITS OF LENGTH

    Micron. One millionth of a meter, or about 39 millionths of an inch; a unit used in expressing the wavelength of electromagnetic radiation.

PART ONE

ENERGY IN NATURE
AND HISTORY

1

HOW THE PROBLEM

CAME TO BE

If you can look into the seeds of time and say which grain will grow and which will not, speak then to me.

Shakespeare

In the beginning was energy. Through transformations of energy, the universe is evolving, ourselves with it. Whenever anything happens in the universe — anything, anywhere — it is because energy has been transferred from one form to another or from something to something else. Energy makes things happen.

In our speck of the universe, we have energy from the molten interior of the earth, from interactions between earth and other celestial bodies, and most importantly, from the sun. In addition to the familiar cycles of heat and light, the effect of sun on the earth is seen in the winds, the cycle of rainfall and evaporation, the tides, in the energy stored in green plants and, several transformations later, in our own bodies.

When our planet was young, radiation from the sun and from space bombarded a primitive atmosphere, providing energy for the formation of simple organic molecules. These molecules were washed into the growing seas by violent rainstorms. In the lightning that accompanied the rain, still more molecules were born. Eventually, some of the molecules came together into groups, and surface films formed around them. Laboratory experiments simulating postulated conditions on the primitive earth show how this might have happened. Something inherent in the geometry of the atoms and molecules and their distribution of energy makes them behave this way. Thus, a tiny fraction of the radiant energy reaching the earth was stored as chemical energy in associations of organic molecules.

Nature's early molecular experiments were random, but nature has never been conservative with materials. From uncounted failures arose some configurations that grew and reproduced themselves. Improbable? Not really. Given enough chances the improbable will happen. We know that at some time in the past, aggregates of molecules passed through the shadow dividing life from non-life.

Probably these first life forms, 2 or 3 billion years ago, grew, multiplied, and derived their energy at the expense of complex molecules that swam in the "primordial soup." Gradually, however, the large molecules were used up, so that only those organisms could survive that were able to synthesize their own complex molecules from simple ones. Perhaps only a few organisms out of billions achieved the necessary autonomy for the survival of themselves and their descendants. Perhaps — and we will never know — it was only one.

Life's invention of photosynthesis, an energy conversion process, was a key point in evolution. It permitted development of the never-ending variety of life forms that have paraded across the earth in the last 3 billion years. Life is energy conversion.

With photosynthesis came the potential for altering the earth's energy balance ever so slightly and for changing the chemical composition of the atmosphere and the outer skin of the earth itself. Molecules of chlorophyll trapped photons of light, becoming activated and initiating a chain of

This volcano on the coast of Alaska is evidence of the energy stored in the earth's interior. (Photo courtesty of Marine Studies Center, University of Wisconsin.)

events leading to the formation of carbohydrates from carbon dioxide and water. Carbohydrates became the fuel of most living cells.

With photosynthesis, larger amounts of solar energy could be diverted to the use of living creatures. For every molecule of carbon dioxide fixed into fuel, a molecule of oxygen was released. Gradually, an oxygen-rich atmosphere evolved through photosynthesis, the photodecomposition of water, and perhaps through geological and geochemical processes. The new atmosphere permitted the evolution of aerobic respiration, a process much more efficient than anaerobic fermentation in utilizing the stored energy of carbohydrate molecules. In the wake of green plants, animals evolved — first in the water, then in pursuit of plants that invaded the land. With the exception of certain chemosynthetic bacteria, even microorganisms, as we know them today, are indebted to photosynthetic plants for their existence.

At times the balance must have been delicate between the green energy trappers and the energy eaters. Had the green plants been depleted, the story of life might have ended. But they continued to multiply. They produced enough not only for their own growth and multiplication, but a surplus so large that not all the energy eaters together, from the microscopic decomposers to the mighty dinosaurs, could consume it. The earth began storing unused organic matter in the form of coal, oil, and oil derivatives — the fossil fuels. It would prove to be capital laid away for exploitation by an insatiable energy eater of the future, a creature yet so remote that it was only hinted at by tiny rodent-like beings that scurried around beneath the feet of dinosaurs.

The dependence of energy eaters on green energy trappers is responsible for the pyramidal form of the food web. Approximately 10 pounds of food are required to produce 1 pound of the creature that is fed. Thus, 100 pounds of plants are needed to produce 10 pounds of herbivore which in turn are converted into 1 pound of carnivore. The pyramid is narrowing rapidly. And the carnivores that prey on the carnivores, and the carnivores that prey on them — all are held parsimoniously in check by their interactions with organisms at other levels of the pyramid.

After perhaps several billion years of evolution, modern man emerged. Until a geological yesterday, the earth's energy balance sheet came out about even, perhaps a little on the plus side with small amounts of energy being laid away. Each day the sun shone and green plants built energy-rich molecules, more than enough for their own growth and survival into future generations. Much of the remainder was eaten by herbivores that were eaten by carnivores, and everything that died furnished food for scavengers and a host of decomposers. There was a surplus of food. Some of it, buried in mud or at the ocean bottom or otherwise protected from decomposition, became unavailable until yesterday, when modern man discovered fossil fuel.

Only man is able to utilize energy in excess of that necessary to satisfy his nutritional needs. His most recent economic and technological development has been at the expense of that capital energy investment, the fossil fuels. Through clever but heedless manipulation of energy, he has brought himself to an energy crisis. We do not yet know the end of the story, but we do know that when capital is used up a common end is bankruptcy. It is possible that a rich uncle may die opportunely and bequeath a source of unlimited capital. It is possible that technology will tame nuclear fusion or solar energy to fill our real and imagined needs — but we may be appalled by the results.

For the first time in history, man has proposed to dig deeply into the vast sources of physical energy that propel the hurricanes and into the nuclear energy that burns the stars. We have no precedent by which to judge what energy conversions of the contemplated magnitude will do to the earth. Yet we shortsightedly concentrate on overcoming technological problems, asking only "Can we do it?", and never "Should we?" We ponder the consequences of our action superficially, if at all. We are teetering precariously at the top of the ecological pyramid. But our increasing numbers and demands seem to make it imperative that we take our place at the bottom of the pyramid, along with green plants, as primary producers. When we scheme to use solar energy directly we become photosynthesizers in the literal meaning of the word. And we can be certain that if man becomes a photosynthesizer there will be profound consequences on earth.

Our global energy problems — lack of food, surplus of waste energy, uncertainty about future energy supply — are a result of human actions. We need more energy, we are told, so that progress can continue. But can more of the behavior that caused the problems solve them? The best that energy technology can do is make things tidier while we struggle to change our habits. And we must change, not only our habits, but some of our most fundamental assumptions and beliefs. We can no longer deny our evolutionary history and our biological and psychological relationship to the earth.

Perhaps you still believe that in our current crisis of crises the energy crisis is less critical than some of the others. We will not run out of fuel for several centuries The population explosion or pollution or nuclear war may do us in before we need to worry about the depletion of fuels, from which technology may yet save us. But we suggest that the population explosion has come about because of our ability to bend energy to our desires, that pollution is caused by the technology to which energy gave rise, that wars are related to inequitable distribution of energy resources (including food), and that the very degree to which war has become a threat is a function of how we have exploited energy to kill one another. Whenever anything happens in the universe, it is because energy has been transferred from something to something else. Energy makes things happen.

As this book goes to press, the United States is facing a critical energy shortage. The immediate cause of the crisis is a cutoff of oil from the Arab nations, and many people assume that after this political problem has been solved the energy problem will also go away. But the current shortage has been in the making for some time, and the crisis was merely precipitated by the Arab situation. The problem will not go away, as this book attempts to show.

The energy crisis is thought by some to be a blessing in disguise. Sooner or later, something will halt our growth in numbers and our consumption of resources. We can name most of the possibilities: war, disease, famine, depletion of resources, pollution of the environment, lack of space. Perhaps the limit imposed by the availability of energy is the kindest of all these. At least we can foresee our future problems, and we can plan for the kind of future we want, so long as it is compatible with the realities of our physical and biological world.

2

ENERGY AND HISTORY: TWO TALES

OF PROMETHEUS

. . . they were running hand in hand, and the Queen went so fast that it was all [Alice] could do to keep up with her; and still the Queen kept crying "Faster! Faster!" The most curious part of the thing was, that . . . however fast they went, they never seemed to pass anything.

". . . in our country," said Alice, . . . "you'd generally get to somewhere else — if you ran very fast for a long time as we've been doing."

"A slow sort of country!" said the Queen. "Now here, you see, it takes all the running you can do, to keep in the same place. If you want to get somewhere else, you must run at least twice as fast as that."

"I'd rather not try, please!" said Alice. "I'm quite content to stay here."

Lewis Carroll

In the beginning man, like other animals, had only his own energy to use in the struggle for life. This came from the food he ate — about 2200 kilocalories* per day averaged over his lifetime, or 843,000 kilocalories per year. That is not very much. It represents a power output about equal to one 100-watt light bulb. Today, many of the world's people still exist at or below this level, for they are starving. But in the United States, per capita

---

* The familiar "food calorie" is equal to 1,000 ordinary calories, or 1 kilocalorie.

energy consumption is roughly 87 million kilocalories per year,† more than one hundred times the energy that operates a human body.

What do we have to show for this incredible use of energy? We convert it to heat, light, motion, sound, and things. We have passed the point of diminishing returns, however, for our per capita energy consumption is rising faster than the standard of living, and our life expectancy and other tangible measures of well-being are below those of some less consumptive societies. This is partly because it takes an increasing amount of energy just to stay where we are. Alice's lament seems so sensible, compared to the obsessive demands of the Queen. Why are we running? Where are we trying to go? Are we unable to see where we are really headed?*

We did not always run in order to stand still. Look at the road we have traveled in a few tens of thousands of years. In agriculture we have gone from manpower to oxen to tractors. In transportation, human feet and backs yielded first to dogs, then to oxen, pack mules, horses, camels, and elephants, which were sometimes attached to mechanical contraptions; now we have jets and spaceships. The tools of war evolved from stones through gunpowder to hydrogen bombs. In medicine, magic grudgingly gave way to vaccines and antiseptics; today we boast of laser machines and artificial kidneys. Leaves were abandoned when we learned to make paper fans; now we have central air conditioning. Communication began as spontaneous information exchange between people; it was revolutionized by writing, again by the printing press, and most recently by satellite TV. The common denominator is the exploitation of energy.

The history of mankind, and of Western man in particular, could be written in terms of the conquest of energy (table 2-1). In timeless cultures where tradition rules, there has been until recently either no desire for or no way to obtain the extra energy that makes things happen.

Much "progress" is clearly misdirected. But you are physically comfortable and well fed, and through most of time your ancestors were not. They were often cold and fearful. They had miscellaneous parasites and infections and broken bones. Many died at birth or giving birth. Few lived past their reproductive period. Many — of all races, in all lands — were slaves, whether or not they were called by that name. That is why they invented Paradise. And no wonder, when they began to see what energy could do for them, that they believed energy would bring Paradise to earth if only they could tame enough of it!

---

† This number is higher than the number sometimes quoted, because it represents primary sources of energy: wood, water, coal, petroleum, nuclear energy.

* In economic terms, the marginal benefits from equal increments of output are decreasing and the marginal costs of equal increments of output are increasing. At some point, extra output costs more than it is worth.

Table 2-1.   Chronology of Events in the History of Man and Energy

B.C.	
1,700,000	First Ice Age begins. Several varieties of erect, man-like primates exist.
Before 500,000	Man begins to use fire.
Before 13,000	Domestication of the dog.
9,000	Beginnings of agriculture.
8,000	Last retreat of the continental ice sheet.
7,000	First sickles, found in Palestine.
6,000	Domestication of goats, pigs, sheep, cattle, oxen.
4,000	Domestication of the horse.
3,500	Wheel invented, probably in Mesopotamia.
3,000	Man learns to smelt metal and make bronze.
1,000	Beginning of iron technology. Domestication of the camel.
300	Waterwheels in Greece.
200	Modern harness invented in China.
100	Horizontal shaft waterwheel produces 0.3 kilowatts power.
27	Book by Vitruvius describes watermills, steam jets, and machines in general.
A.D.	
300	Vertical shaft waterwheel produces 2 kilowatts power.
500	Waterwheels come to Europe.
650	First windmills. Modern horse harness reinvented in Europe.
852	Coal burned in an English monastery.
900	Whale oil used for lighting.
1239	Coal used as fuel by smiths and brewers.
1300	First coal used in home heating.
1404	First giant cannon, in Austria.
1500	Tide mills in the Netherlands; windmills used to drain submerged lands and maintain them.
1600	Versailles water works produces 56 kilowatts power.
1606	First known experimental steam engine built by Della Porta.
1673	Huygens builds internal combustion engine run on gunpowder.
1690	Papin designs the first piston engine.
1693	Leibnitz states the law of conservation of potential and kinetic energy.
1712	Newcomen builds first steam pumping engine.
1740	Improvements in iron technology.
1765	Modern steam engine conceived by Watt.
1789	Coulomb's work in electrostatics.
1820-1860	Work of Oersted, Ampere, Faraday, and Maxwell in electricity. Principles of thermodynamics worked out by Carnot and Clausius.
1859	First oil well drilled in Pennsylvania.
1866	Transatlantic cable laid.
1876	Otto designs four-stroke internal combustion engine.
1882	First incandescent lighting, New York.
1895	Roentgen discovers X-rays.
1896	Becquerel discovers radioactivity.
1898	Tsiolkovski works out principles of rocket flight.
1903	First flight of Wright brothers.
1920	First scheduled public radio broadcast, Pittsburgh.
1926	Goddard fires first rockets with liquid propellant.
1941	First jet plane flight.
1942	Fermi starts first atomic reactor in Chicago.
1945	First nuclear explosion, Alamogordo, New Mexico.

Table 2.1 continued

1945 Uranium bomb dropped on Hiroshima; plutonium bomb, on Nagasaki.	service near Moscow, U.S.S.R.
1952 First hydrogen bomb explosion.	1957 Sputnik I, first artificial satellite. First nuclear power plant in U.S., Shippingport, Pennsylvania.
1954 Launching of first atomic-powered submarine, *Nautilus*. First nuclear power plant put into	1969 Man's historic landing on the moon.

Now, man has walked on the moon. Artificial moons gather data about the weather and beam the latest news onto television screens in India and America, alike. Nations vie with one another to build faster-than-sound aircraft to carry businessmen from breakfast in Moscow to lunch in London.

The almost 4 billion people of the world today use energy at a rate of about 15.7 billion kilowatts, not including food. With the population approaching 4 billion, this is about 1.4 kilowatts per person. If food is included, every person in the world can be imagined as radiating the energy of fifteen 100-watt light bulbs. In the United States, about 200 million people glow at the per capita rate of one hundred and sixteen 100-watt light bulbs.

Forty or fifty years ago, in contrast, world population was less than 2 billion. Each of these people was using energy at a rate of about ten 100-watt light bulbs. The United States even then led the pack in power consumption by a 10 to 1 margin. Radio was revolutionizing home entertainment. A war which killed 5.5 million people seemed so terrible that another war was unthinkable. The number of cars in America had grown to more than 8 million. One out of eight adults now had an automobile which carried him bumpily and uncertainly over rudimentary highways. On the farms, mules and horses outnumbered tractors 45 to 1. One-quarter of America's farmland was planted in crops to feed her 25 million draft animals. In the 1920s most homes were heated with coal. Wood was on the way out and petroleum products were on the way in. The first electric refrigerators were beginning to displace the ice box, and electricity was already supplying substantial power to industry.

If we go back a few more decades, to the end of the nineteenth century, we will find that the world was darker then. Homes could not be flooded with light to make midnight rival noon. The first oil booms, which led to a frantic search for new markets, had permitted petroleum products to replace scarce and expensive whale oil for illumination. On September 4, 1882, the Pearl Street Station in Manhattan first switched on to furnish power for four hundred light bulbs of 83 watts each. It was a sign of things to come.

A peak of modern energy intensity and energy dependence was reached when Buzz Aldrin arrived on the moon. (Photo courtesy of NASA.)

After a three-hundred-year gestation period, a practical internal combustion engine had been born. The 1890s witnessed the first cars in both Europe and America and the first flying machine with an internal combustion engine — a fatal experiment for its inventor. Further hastened by the evolution of electric motors, the end of the steam engine was in sight even as it was achieving its most remarkable feats.

Animals provided man's extra power for agriculture, as well as much of his transportation. Wood heated his home and cooked his food. Coal, wind, and water — but by now mostly coal — drove his industrial machines. There were 1.5 billion people in the world using power at the rate of about 700 watts per person. Becquerel discovered radioactivity in 1896, but it would be half a century before people would comprehend the full meaning of his discovery.

During the eighteenth and nineteenth centuries, the Western world was convulsed by the Industrial Revolution, while the rest of the world went on much as it had for thousands of years. The Industrial Revolution came late to America, sparing her some of Europe's agony. The new continent offered abundant opportunities in agriculture and a challenge to exploration. Population density was low, distances enormous, and transportation primitive. The predominant pattern in the colonial and early national period was one of skilled home and local industries, run by master craftsmen with a few apprentices who themselves hoped to be master craftsmen one day. All but the newest and most isolated communities had mechanical power, chiefly water, sometimes wind, for specialized heavy tasks. In fact, the mill often came first, and the town huddled around it.

In 1850, gristmills like this one were found in almost every town with a suitable stream. (Photo courtesy of U.S. Department of Agriculture.)

Riverside gristmills and sawmills were small but scattered thickly across the land, for a round trip of 25 miles to the mill and back was as much as most farmers could negotiate in a day. Local iron works, paper mills, brickyards, breweries, and other small industries were common. Communities had to be self-sufficient.

In the last half of the nineteenth century, many things led to the industrial development of America — creation of vast private fortunes; new waves of immigration providing, for the first time, abundant labor; the development of marketing techniques and the market system; the expanding network of railroads; and a burst of inventions. With energy from the steam engine for transportation and heat-requiring industries, and energy from further refinement of the waterwheel and the first hydraulic turbines for manufacturing, craft after craft gave way to mass production.

In 1850, energy consumption in the United States was already at a rate of 4.5 kilowatts per capita, but the average for the world's billion people was only 0.4 kilowatts. Of the energy used in the United States, two-thirds was supplied by draft animals and by man himself. Of the remaining third, two-thirds was supplied by wind and water and one-third by fuels. Ninety percent of this fuel was wood; 10 percent, coal. The first oil well was yet to be drilled. In Europe the story was much the same, except that some countries burned more coal because they had depleted their forests.

The Industrial Revolution was born in England in the eighteenth century, amidst a flurry of mechanical inventions and technological advances whose names are familiar and will not be documented here. Of the inventions, the steam engine was king. Improved iron and steel technology stood behind much of the industrial development.

Social and economic changes had been preparing the way for the revolution for five hundred years. Most people bitterly opposed the change imposed by the Industrial Revolution. Innovation had been fought for centuries, by individuals, craft and trade guilds, and govenments. New patent applications were routinely denied when they threatened old ways, and people were put to death for defying convention. But invention became an obsession. Even when temporarily defeated, it rose again like the phoenix, stronger than before.

Eighteenth-century England reeled under rampant poverty and unemployment, resulting largely from an unprecedented land grab (which also took place on the continent) begun by sheep-raising noblemen a hundred years before. It threw many tens of thousands of agricultural workers off what had been the common land. At first, a dismayed government restricted the paupers to their local parishes, on poor relief. Later, when factories grew, paupers flocked to the cities, even as the agricultural poor do today. All over England, humanity rebelled against the tyranny of the machine and against the conditions of crowding and poverty that were laid

Early in the Industrial Revolution, factories and the village houses huddled by the water's edge. Factories belched black smoke, but populations were smaller and air pollution was a local affair. (Photo courtesy of Wisconsin Historical Society.)

to it. The situation was very like that of recent years, with the affluent calling for law and order, feeling that not only their security but their lives were in danger. The mobs of rioting poor were protesting their alienation as much as their poverty. Rumor spread that a General Ludd or King Ludd was masterminding the rebellion. His followers, the Luddites, wanted to blot out all social, economic, and technological change and to return to some mythical utopia of the past. They were not successful. History cannot be erased and rewritten; time cannot be reversed. Nothing can be accomplished unless we accept the present, and begin from there.

Lewis Mumford argues persuasively that the technological ingredients of the Industrial Revolution had lain around for centuries, waiting to be blended with the right social, political, and economic conditions in the eighteenth century.[1] Man had mastered the principles of machines a thousand years before the Industrial Revolution and could use wheels, gears, levers, and pulleys to his advantage. Prototype machines of many descriptions had been invented by the Egyptians, the Chinese, the Arabs, the Cretans, and later the Greeks and Romans. Many of these machines seemed to be but toys, created for amusement.

There was also plenty of power. Despite the excitement about the steam engine, the early Industrial Revolution was based on waterwheels and windmills. The first windmills had appeared in the Moslem world about A.D. 650, they were used in China from the year 1000 on, and by the fourteenth century they had spread throughout Europe. The gigantic land reclamation projects of the Netherlands were accomplished by means of windpower during the sixteenth century. Waterwheels had been in existence still longer. They probably originated in the Near East in the fourth or third century B.C. and spread through Rome to Europe. By the twelfth century there were thousands of waterwheels throughout western Europe. It is difficult to estimate the total power that was available at any particular time, but we do know the approximate power output of the most common machines. Table 2.2 shows the power available from basic machines, from primitive to contemporary times.

With power and machines widely available in the Middle Ages, why did the Industrial Revolution wait until 1700 to begin? Mumford believes that the probable explanation involves a breakdown in European society, the substitution of competition and personal gain for the medieval traditions that had prevailed, the substitution of abstract values for life values, and preoccupation with insignificant luxuries. Certainly social and economic conditions were at the root of the change. Technically, it could have begun centuries earlier. It could have begun in China or somewhere else. But it began in England about 1700 and still reverberates around the world, and we are left to deal with the consequences as best we can.

Heralds of the Revolution included both technological optimists and technological pessimists, much as our forecasters do today. Environmental

Table 2-2.   Power Output of Some Basic Machines

A prime mover is a machine which converts food, fuel, or force to work or power. Because power is the rate at which energy flows, a machine with a low power rating can often do the same job as one with a high power rating, but it takes a longer time. The energy in a gallon of gasoline can operate your car for half an hour with a power output of 38 kilowatts. It would run a jet airplane, at 45,000 kilowatts, for a second and a half. Listed in order of increasing power output, the machines below are also in approximate order of their invention or exploitation by man. The power output of prime movers has increased 100 million times in the last ten thousand years, from the power of man and his domestic animals to that of a liquid fuel rocket.

Prime mover	Typical Power Output (kilowatts)
Man	0.1
Ox	0.2
Horse	0.5
Windmill	15
Waterwheel	3000
Steam engine	2,000
Internal combustion engine	10,000
Gas turbine	80,000
Water turbine	100,000
Steam turbine	1,000,000
Liquid fuel rocket	16,000,000

problems and the bleak realities of life contrasted shockingly with what people thought could and should be. Here is a commentary on mining, written early in the sixteenth century:

> The critics say further that mining is a perilous occupation to pursue because the miners are sometimes killed by the pestilential air which they breathe; sometimes their lungs rot away; sometimes the men perish by being crushed in masses of rock; sometimes falling from ladders into the shafts, they break their arms, legs, or necks. . . . Besides this the strongest argument of the detractors is that the fields are devastated by mining operations, for which reason formerly Italians were warned by law that no one should dig the earth for metals and so injure their very fertile fields, their vineyards, and their olive groves. Also they argue that the woods and groves are cut down, for there is need for endless amount of wood for timbers, machines, and the smelting of metals. And when the woods and groves are felled, there are exterminate the beasts and birds. . . . Further, when the ores are washed, the water which has been used poisons the brooks and streams, and either destroys the fish or drives them away.[2]

Contrast this prophecy, written in the seventeenth century:

> I doubt not posterity will find many things that are now but rumors verified into practical realities. It may be that, some ages hence, a voyage to the Southern tracts, yea, possibly to the moon, will not be more strange than one to America. . . . The restoration of grey hairs to juvenility and the renewing the exhausted marrow may at length be effected without a miracle; and the turning of the now comparatively desert world into a paradise may not improbably be effected from late agriculture.[3]

Led by Francis Bacon, the greatest thinkers of the Renaissance and early seventeenth century wrote a prescription for the course that science and technology would pursue into the twenty-first century. They wrote it at a time when technology in the modern sense had barely been conceived, when scientific imagination was hundreds of years ahead of technical ability. Yet the predictions were accurate and the prophecies self-fulfilling, except for one detail: the place to which energy and the machine would lead man was supposed to be utopia.

One of the many reasons that utopia eluded us was our inclination toward warfare. Man has always pursued the technology of weapons with vigor — the first tools were also weapons. Developments in energy use, iron technology, and explosives were united in Austria in 1404 to produce the first giant cannon on record. It weighed almost 5 tons. In 1450, near Caen, in Normandy, four thousand Frenchmen and two cannons faced seven thousand Englishmen and demolished them. The body count: twelve French, fifty-six hundred English. Two years later, with cannon, the Turks conquered Constantinople.

France built her first blast furnaces around 1500. By 1600, there were thirteen foundries in France, all of them manufacturing cannons. Sweden and Russia also had arms factories. When the sewing machine was invented in 1829, the French War Department was the first to try it, for mass producing uniforms. An army is a wonderful thing for the economy.

What of life before Europe's surrender to the machine? History, as it is often written, suggests that the Middle Ages — the period between A.D. 400 and 1400 (1300, in Italy) — were a time of dismal stagnation. Curiosity was absent, if not dead, and there was no concern with progress. After this deep and dreamless sleep, people awoke in the bright morning of the Renaissance and shook themselves back to reality. One of the major events that occurred in the dawn was the rise of capitalism. It is interesting to ponder whether the goals and abstractions of capitalism were more real as a guide for life than the saints, angels, and demons that cast their shadows on the medieval world. Capitalism was the ultimate alchemy. It transmuted everything into gold.

The energy of the Middle Ages did not grossly pollute air, soil, and water. Wind and water power were clean and free; wood, if judiciously managed, was replaceable; and animals were not yet in competition with man for food. Although life was not easy, men of the Middle Ages created a panorama of farms, towns, cities, gardens, canals, and harbors that in many ways shamed the squalid scenes to follow. The beauty of the towns and country contrasted cruelly, however, with the primitive medical and sanitary practices that led to periodic pandemics of the Black Death, in which an estimated quarter of the population perished.

It is difficult to estimate how much power was available during the Middle Ages. The use of power was low, only a few hundred watts per person. There was a literal increase in horsepower, made possible by the invention of iron horseshoes and the modern harness. (The Chinese had this harness in 200 B.C.; Europeans reinvented it eleven hundred years later.) Another new source of energy was coal. The earliest recorded use of coal as a fuel was at the Abbey of Peterborough, England, in 852. From that obscure time when monks first gathered "sea coales" from England's rocky shores, the demand for coal would grow until, at the turn of the twentieth century, coal would become the number one source of energy in the world. In 852, sea coales were mostly a curiosity. Two hundred years later they had begun to take pressure off Europe's vanishing forests, and through their mining and use man embarked on a new phase of the exploitation of his planet.

Water power was rapidly developed along the Rhine, the Danube, Italy's swift streams, and in North Sea and Baltic lands. Water mills ground grain, sawed wood, pulped rags for paper, pumped water, crushed ore and otherwise furthered mining and metalworking, and fostered the development of Europe's textile industry. Wind was more difficult to harness, although where conditions were favorable it was the main source of power. Wind had driven sailboats since Stone Age times. The close of the Middle Ages saw the opening of an era of sailing and discovery that lasted for hundreds of years, an age that came to a dramatic end in 1929 when the oceanographic research vessel *Carnegie* burned in Pago Pago.

The technology of the Middle Ages depended on the smelting and working of iron, with wood for fuel. But the foundations of mining and metallurgy had been laid thousands of years before, when man's glance was first arrested by the glitter of gold. In prehistory men found chunks of native metal and pounded them into various utilitarian objects. But they could not get very far with this technique. It was difficult to find metal in useful amounts, and there was a limit to what could be accomplished by brute force pounding. Sometime around 3000 B.C. came a dramatic moment when someone observed that pure liquid melted from certain rocks when they were heated, and this metal was much easier to work than that obtained in the old way. Thus, the period commonly called the Bronze Age signified the discovery of a brand new use of energy and the beginning of mining and metalworking.

Copper, silver, gold, and meteoritic iron were found in their native metallic state in prehistoric times, but man knew other metals as well, including lead and tin. By accident or experiment, man learned to smelt these ores in combination with charred wood, which reduced the chemically combined metal to its elemental state. He learned to fracture rocks by building fires against them and splashing them with cold water. He added a new dimension to his exploitation of energy and the earth.

Before metalworking began, the only division of labor was according to age and sex. In the later Stone Age, it is likely that certain people, perhaps the older men, produced many of the common flint tools and taught their art to others. But anyone could make his own ax or knife or arrowhead if he had to. Not so, with the strenuous and exacting art of the smith. Smiths were probably the first specialized craftsmen, and their appearance marks the beginning of far-reaching social change. The change was resisted, as change always is. Stone tools persisted in Egyptian agriculture for two thousand years after the superiority of metal products had been demonstrated. This may be partly a sign of the failure of supply to meet demand, but it must also result from the farmer's balkiness in accepting innovation.

There were about 100 million people in the eleventh century B.C., a century that witnessed the entry of iron, camels, large riding horses, dwindling forests, soil erosion, simplified writing, and coined money into the affairs of men. Camels and goats gave man mobility and high quality protein. They permitted him to live in areas where he could not support himself by agriculture. They also girdled trees and bushes and cropped the grazing lands too close and helped to expand the deserts.

The price of improvement everywhere was increasing degradation of the earth's resources. The result of improvement was specialization and growth in the size and complexity of social institutions. Energy was increasingly diverted away from life support and into the building and maintenance of social structure. And when society began to grow complex, man domesticated yet another animal to provide more energy: man. Slaves built the pyramid of Cheops and slaves supported Greek civilization.

Individual hand technology and simple human relations could not give way to specialization, group production, machines, and a complicated social structure until centers of high population density were formed. As long as man obtained his food by hunting and gathering, no permanent social grouping larger than an extended family or tribe was possible because so much territory was needed to supply each person's food. Without agriculture or industry, the earth might support between 20 and 40 million people — hunting, fishing, collecting fruits and nuts, digging for roots, and catching whatever worms, insects, frogs, snakes, and turtles they could find. At present, instead of 40 million, our numbers are rapidly approaching one hundred times that figure. Only one thing makes this possible: agriculture.

At the dawn of agriculture there were about 5 million people on the earth, spread thinly over the 20 million square miles that were hospitable enough to support them. Observations of hunting and gathering cultures remaining today suggest a maximum population density of one person per square mile at this level of culture. So although the earth was not crowded ten thousand years ago, the population may have been about a quarter of that which could be supported.

In most hunting and gathering cultures, men hunted and women gathered. Modern studies of primitive peoples suggest that woman's labors provided most of the food except in the far north, where people had to be almost totally carnivorous. The hunt yielded a little high quality protein and a lot of prestige. As the women gathered fruits and seeds, they sometimes scattered some of them in the hope of producing more of the favored foods. When this was successful, agriculture began.

Agriculture was first practiced between nine and eleven thousand years ago in the "fertile crescent" of the Middle East, the area bordering what is now Iraq and Iran. The transition from hunting-gathering to agriculture represented a transition from one energy plateau to another. It was an enormous step. The taming of the plant allowed man to appropriate an increased portion of the biosphere's energy for himself and to concentrate himself in towns and cities where people could devote themselves to pursuits other than finding food.

About 7000 B.C., man began burning vast areas of grassland and forest in order to plant crops. This had various effects, depending on the nature of the soil and climate. Burning did not hurt the rich volcanic slopes of Costa Rica, which have remained fertile for thousands of years. Since the beginning of agriculture, however, some areas have been rendered useless because man tried to take more energy out of the system than he put into it.

When man turned from hunting to farming, he tapped the reserves of soil whose fertility had accrued through unmolested generations of plant life and death and the integrated activity of animals and microorganisms. Man burned, planted, and harvested; soon the investment of the years was used up and he had to move on. By the time of Christ there were between 200 and 300 million people in the world, and environmental problems were already critical in many places. The per capita use of power remained low: about 100 watts for food and another 100 for fire and animals. But the number of men was increasing. That made all the difference.

During the half million years before the invention of agriculture, man discovered many uses of fire, invented the bow, further refined stone tools, invented or improved clothing, devised primitive means of land and water transportation, and domesticated the dog. All these inventions had one thing in common: by increasing the fraction of energy available to man, they tipped the ecologic balance in favor of an increased human population and forced corresponding adjustments in all other populations.

Fire kept man warm. It cooked his food, protected him from predators, and drove game toward hunters waiting in ambush. It cleared underbrush from the forest to make hunting easier, and it burned the brush from grasslands to maintain the grasslands and improve the game. At first, man took advantage of natural fires for these purposes. Later, he learned to start his own fire from zealously guarded coals or a slow-burning log. Using fire by no means implied knowing how to make it. The art of making fire was mastered relatively late — around 6000 B.C. — and even then, not by all cultures.

Between 35,000 B.C. and 7000 B.C., man expanded over the face of the globe. It was not an exploratory spirit so much as the quest for game that led him through Siberia, over the land bridge across the Bering Strait, into North America, and then eastward across the continent and southward to the tip of South America. During this time, he tamed his second source of energy: the wolf.

Man and wolf-dog became hunting partners. Wolf-dog extended man's senses and man enhanced wolf-dog's efficiency in the kill. Man induced wolf-dog to pull sleds of stone, skin, or wood. Now, with fire and dog, man could survive in the northern lands wherever there was game.

Just as wolves were probably tamed many times by different groups of people, so was fire. How many times did man gain fire only to lose it again? Fire mattered most where it was cold — where man cherished fire as he cherished life. Many anthropologists believe that small bands of hunters, pursuing game in the snowy north, frequently lost their fire and died. Accidents like this may have contributed to the disappearance of Neanderthal Man.

Half a million years ago there were several varieties of erect, tool-using primates who were men or almost men. Probably some of them could speak. For more than a million years, most of them had lived where the climate was warm and friendly. This was the time of the Pleistocene, when the mild earth grew inhospitable and great walls of ice crept imperceptibly over most of Europe and North America, only to retreat and then advance again. Between the glaciers, man began to extend his range northward, following herds of game which followed the edge of the melting glacier. Many species of mammals met extinction during the Pleistocene epoch. Of the several types of man entering the Ice Age, only one, *Homo sapiens*, survived. With help from nature, he annihilated or assimilated all the rest.

Man had no physical advantage over other animals. His simple tools merely gave him an equal chance, for he was slow-footed, weak, and without claws or fangs. Most men feared the fire that lightning brought and ran from it, with the other animals. But now and then in the cold regions, a man would pause to look at a burning log and wonder dimly. The man would watch the log uncertainly, curiously, and then warm his hands before moving on. Perhaps fire need not be wild and threatening, perhaps he could use it. Half a million years ago, near what is now Peking *Homo erectus Pekinensis* rolled a glowing log to the mouth of his cave.

According to Greek legend Zeus did not intend that man have fire, but Prometheus stole it from heaven and gave it to him. The sequel is not clear. According to Hesiod (ca. 700 B.C.), Zeus threw a tantrum and ordered the creation of Pandora, who came to earth and released her great jarful of evils so that war, violence, greed, and all their accompanying plagues spread over the land. Thus, the gift of Prometheus became man's everlasting curse. Aeschylus, however, exalts Prometheus as the bringer not only of fire but through it survival, along with all the arts and sciences, and everything human and good. We need not take one side or the other. It is both ways.

References

1. L. Mumford, *Technics and Civilization* (New York: Harcourt, Brace, and Company, 1934), p. 184.
2. *Ibid*, p. 71.
3. *Ibid*, p. 59.

3

LIMITS: EFFICIENCY AND GROWTH

*All our drive and optimism are bound up with continuous growth;
"growth addiction" is the unwritten and unconfessed religion of
our times. In industry and also for nations, growth has become
synonymous with hope. Undoubtedly, quantitative growth will
have to go on for many more years, but unless we prepare for a
turning-point well before the end of the century, it may by then be
too late. . . . The insane quantitative growth must stop; but innovation must not stop — it must take an entirely new direction.*

Dennis Gabor

Energy and matter are natural resources. Eager to exploit material
resources, we have often overlooked the role of energy or the fact that
energy resources are limited as surely as are copper and gold. We forget
that only energy makes things happen. In the days when men struggled to
realize grandiose dreams with primitive machines and slave labor, they understood that energy was a precious commodity. But today in the United
States we try to deny our limitations and talk of going to the stars. At the
same time we curse the bungling power company for brownouts,
blackouts, and pollution.

We are using up our energy supplies as surely as we are using up our
metals. In the case of metals, it eventually becomes technically,
economically, and energetically infeasible to reclaim them from wastes or
to mine lower grade ores. As for energy, the inexorable laws of thermodynamics take over as we perpetually degrade concentrated energy to

Growing populations and even faster growth in energy use can bring crowding of a sort we have rarely experienced before. On a nice day at Jones Beach, both the beach and the parking lot are full. (Photo courtesy of Long Island State Park and Recreation Commission.)

diffuse, useless heat. The laws of thermodynamics can be paraphrased to make their message profoundly clear: You can't win; you can't break even; you can't get out of the game. Or, you can never get more out of a system than you put into it. In fact, you can't even get as much. And although according to theory all systems are running down, nothing will ever come to a complete stop.

Nature cycles materials again and again, but energy flows through a system only once and in one invariable direction: toward dissipation as heat. If on first glance it appears that this principle has been violated, a second look will reveal that additional energy has been supplied to the system. The cycles of nature depend on a steady supply of solar energy. An understanding of material cycles and energy flow in an ecosystem is of fundamental importance in ecology. Similarly, we must appreciate the energy requirements of our own attempts at recycling materials. The principle of the non-cycling of energy is implicit in much of the discussion throughout this book.

Energy exists in a variety of forms which are rather freely interconvertible. Different types of energy have customarily been expressed in different units. This makes their equivalence and their relationships difficult to comprehend. In this book we have used the unit watt in discussing power and the kilowatt-hour or the calorie for energy. Energy is the quantity that changes form when work is done or when something happens, while power is the rate at which energy does work. Some examples of the conversion of energy from one form to another are shown in table 3-1.

Energy is required to build and maintain all complex and organized structures, from mountain ranges to living organisms. Spontaneously, order shakes itself into disorder. The mountain range lifted up through vast expenditures of the energy of the earth is slowly eroded flat. Organisms die, and their complex molecules return to simpler form. When Humpty Dumpty converted his potential energy of position first into kinetic energy of falling and then into the heat and sound of the impact and the mechanical energy of scrambling, he was pursuing the inevitable course that all things take. Although in theory it is possible to restore Humpty to his original condition through appropriate inputs of energy, in practice there is no way to unscramble a scrambled egg.

We want to get as much return as possible for our use of energy; we want to derive the maximum useful work from the energy we use. The degree to which we achieve this goal measures our *efficiency*, which is defined as the amount of useful work (or energy output) divided by the amount of energy input. Every process has a maximum theoretical efficiency which puts an upper limit on our expectations and an actual efficiency, lower than the theoretical, which we realize with our imperfect systems. The maximum value for efficiency is 1 (or 100 percent), because the output can never exceed the input.

The study of thermodynamics was born from the discovery of the equivalence between heat and work — in other words, the discovery that heat is a form of energy in action. Heat can be converted to mechanical energy by allowing it to flow from something warm to something cooler (never the reverse) through a mechanism that diverts part of it on the way. This is the principle of a heat engine, of which the steam engine is a familiar example. The possibility of accomplishing useful work by taking advantage of the uneven distribution of heat over the earth has intrigued inventors. Could we, for example, extract heat from the oceans? Theoretically, we could couple the ocean's warm surface waters with the colder regions below, obtaining work in the process. But here again the laws of thermodynamics frustrate us. The amount of work which heat can do is determined by the initial and final temperatures of the system — in other words, by the temperature differential that exists between the warm body and the cooler one. In the case of the oceans, the temperatures involved are relatively low and the temperature differences small. Although a very large quantity of heat is stored in the oceans, it is of low intensity (as measured by the water's temperature). In practice, it does not pay to run a heat engine unless the initial temperature is quite high and the temperature differential between the warmer and the cooler body quite large.

You will hear much about efforts to increase the efficiency of various

Table 3-1. Some Examples of Energy Conversion and Energy Converting Devices

	From					
To	Mechanical	Thermal	Acoustical	Chemical	Electrical	Light
Mechanical	Oar Sail Jack Bicycle	Steam engine	Barograph Ear	Muscle contraction Bomb Jet engine	Electric motor Piezo-electric crystal	Photoelectric door opener
Thermal	Friction Brake Heat pump	Radiator	Sound absorber	Food Fuel Match	Resistor Spark plug	Solar cooker Greenhouse effect
Acoustical	Bell Violin Wind-up phonograph	Flame tube	Megaphone	Explosion	Telephone receiver Loud speaker Thunder	
Chemical	Impact detonation of nitroglycerine	Endothermic chemical reactions		Growth and metabolism	Electrolysis	Photosynthesis Photochemical reactions
Electrical	Dynamo Piezo-electric crystal	Thermopile	Induction microphone	Battery Fuel cell	Transformer	Solar cell
Light	Friction (sparks)	Thermoluminescence	Rock music light shows	Bioluminescence Candle	Lightbulb Lightning	Fluorescence

energy conversion processes. These efforts are motivated by two considerations. First, we strive for efficiency as a conservation measure, to make the most of dwindling energy resources. As the following discussion will show, however, improvements in efficiency cannot in the long term solve the world's energy problem. The second reason for improving efficiency, which may be even more compelling than the first, is that wasted energy is the source of much pollution. When only a small fraction of the energy we use is converted to useful work, the remainder does not disappear. It is still very much with us, and it returns to haunt us in many ways. Since we have little control over this wasted energy, we must try to channel as much as possible into processes that are understood and controllable. With this in mind, we will look at past gains in the efficiency of energy conversion and at the prospects for the future.

To some degree, each of us is aware of the efficiency of an energy conversion process. For fuel burned in an automobile or power plant, this awareness corresponds to knowledge of the physical calculation of efficiency. But for many processes, our knowledge is less precise. If we build a fire to keep warm, how do we gauge its efficiency? By whether it warms us, our feelings say. But then we would have to consider how many people were kept warm and at what temperature and distance from the fire. The actual calculation of physical efficiency for a campfire warming four people at a distance of 4 feet would give a value of at most a few percent; but it is unlikely that this number, even if known, would govern our actions in building or feeding the fire. How difficult it was to find fuel and the smokiness of the fire would be more important.

The calculation of efficiency depends on the point of view. For example, if we want to know about the process of transporting the automobile over some fixed distance, we must include in our calculation the nature of the terrain and the mass, including the load, of the automobile. If we want to ask the larger social question of the automobile's efficiency as a means of transporting people, then we must include the number of people in our calculation and obtain figures in terms of passenger miles per gallon of gasoline.

These larger considerations, which are often crucially important in questions of energy supply and demand, depend, however, on the basic efficiency of the energy conversion process — from fuel to work or from one kind of energy to another. Thus, in what follows we will first examine the problem of the efficiency of energy converters. To the extent that we obtain gains in efficiency, our fuel requirements are eased and the waste heat which may constitute an environmental hazard is diminished. We will return later to some comments on the larger considerations.

The Industrial Revolution ushered in a period in which the overall efficiency of fuel use (whether food or coal) actually declined. The thumping, chugging steam engines of early industrial plants were enormously wasteful machines. James Watt belied his Scottish ancestry in producing a steam engine which wasted more than 95 percent of the energy of its fuel. Even domestic animals were more efficient in their use of fuel than these early steam engines. Efficiency, of course, was not a prime consideration then. Sheer power and independence from the vagaries of falling water, wind, and tiring animals were far more important. Similarly, the great age of railroads was driven by steam engines with efficiencies of 10 percent or less. The advent of the internal combustion engine represented a real gain, for even the early ones had efficiencies near 20 percent. Early electrical plants were even less efficient than early steam engines. In 1900 the average efficiency of conversion of fuel to electricity was less than 5 percent.

Through the development of new processes or improvements in old ones, there have been large gains in efficiency since 1900. The efficiency of electric power plants now averages about 33 percent, and for the newest plants it is over 40 percent. The change from steam locomotives to diesel power in trains increased efficiency almost sixfold. In terms of actual fuel

Central heating was rare or unknown when this old house was built. The multiple chimneys show that fireplaces and stoves were used to provide heat in the cold New England winters. Even then much of the house was probably quite cold by today's standards. (Photo courtesy of Marine Studies Center, University of Wisconsin.)

use, some of these gains seem quite dramatic. For example, in 1900 it took about 7 pounds of coal to generate a kilowatt-hour of electricity, while in 1960 it took less than a pound. We have also made gains in space heating. From the fireplaces of earlier times to the central oil and gas furnaces of today, we have had about a fourfold increase in the efficiency of fuel use. Efficiency, however, was not the sole consideration in adopting these changed methods. Cleanliness, convenience, and a host of other factors played large roles. Even now, efficiency of fuel use does not govern the choice, for the introduction of electric heat, however convenient, represents an efficiency of fuel use only half that of a modern oil or gas furnace.

A similar situation holds for lighting. The incandescent bulb, which shows no sign of disappearing from use, is an extraordinarily wasteful device: only 5 percent of the electricity is converted into visible light. (Even so, it is about five times more efficient than the carbon filament lamps of 1900.) More efficient forms of lighting are known: fluorescent lights, with an efficiency of about 20 percent; high intensity lamps, reaching efficiencies as high as 35 percent; and various vapor lamps with efficiencies exceeding 30 percent. Clearly, if the cost of energy is not the prime consideration, the choice of lighting, heating, or other conversion processes will not be governed by considerations of efficiency.

Table 3-2 presents the efficiencies of some energy converters in common use. Not all processes show steady improvement in technical efficiency. The internal combustion engine of forty years ago was only slightly less efficient than that in common use today. In the case of automobiles, the increase in horsepower and size without a corresponding increase in the number of passengers has meant that the overall efficiency of automobile transportation has actually declined.

Perhaps the most dramatic gain in industrial processes has come through the introduction of the electric motor. Electric motors can operate at efficiencies ranging between 70 and 90 percent or more, while the old central steam prime mover had an efficiency of 10 percent or less. Since we must allow for the inefficiency of the electrical conversion process, the overall efficiency of electric motors drops to between 25 and 35 percent (the efficiency of generating electricity multiplied by the efficiency of the motor itself). In industrial processes, however, the electric motor's range of sizes and its ability to be turned on and off instantly provide for an efficiency of design which partially offsets the inefficiency of the electric conversion process.

Past gains in the efficiency of energy conversion are of far more than historical interest. If efficiency had remained constant at 1910 levels, the expansion of energy consumption would have been far greater than what we have experienced. Estimates of future demand are often made by projecting past experience. But this implies that improvements in the

Table 3-2.   Technical Efficiency of Energy Conversion

Device	Efficiency of Electrical Conversion (percent)	Efficiency of Primary Fuel Use (percent)
Space heating		
gas furnace	—	75-85
oil furnace	—	60-75
coal furnace	—	55-70
electric resistance heater	95	30-38
Water heating		
gas water heater	—	60-70
oil water heater	—	50-55
electric water heater	90-92	29-37
Cooking		
gas stove	—	60-70
electric stove	75	24-30
Home appliances		
gas clothes dryer	—	45-50
electric clothes dryer	57	18-23
gas refrigerator	—	30
electric refrigerator	50	16-20
gas air conditioner	—	30
electric air conditioner	50	16-20
small motor appliances	50-70	16-28
Lighting		
high intensity lamp	33	11-13
fluorescent lamp	23-28	7-11
incandescent lamp	4	1.3-1.6
mantle lamp (gas or kerosene)	—	0.5
Transportation		
diesel engine	—	38
gas turbine	—	36
automobile engine	—	23-25
Wankel rotary engine	—	18
steam locomotive	—	8
Miscellaneous		
large electric motor	90-95	29-38
large steam boiler	—	88
storage battery (wet)	—	70-75
fuel cell	—	60
steam turbine	—	45
gas laser	38-40	12-16
steam power plant	—	32-40
nuclear power plant	—	32-33
solar cell	—	10

processes of energy conversion will continue at the same rate as in the past. We would like to know, therefore, to what extent past gains in efficiency have held down the increase in demand for energy. We would also like to know the prospects for continued gains in the future.

But there is another reason for our interest in efficiency: the burden of pollution placed on the environment by the conversion of energy. The relationship is not simple except in the case of thermal pollution. In this case, all of the fuel consumed eventually winds up as waste heat, which may be a local or global environmental problem. The more efficiently we employ energy to do work, the smaller the thermal burden on the environment will be. Pollution by chemicals also results from energy conversion. The more fuel we consume, the more chemical pollutants will be released for any given level of pollution control. Since pollution control can never be perfect, one way to minimize chemical pollutants is to consume less fuel.

It is difficult to estimate the overall gains in efficiency for the United States or the world because the estimate depends on different effects in different sectors of energy use. We estimate that efficiency may have increased as much as fivefold in the period since 1900. Another estimate is a fourfold increase.[1] Efficiency appears to have declined somewhat in the period between 1880 and 1920. During this time, relatively inefficient methods were replacing older, more efficient methods for reasons that had little to do with efficiency. The prime gains in efficiency occurred between 1930 and the present. Overall, improvements in efficiency have probably been more than twofold and less than fivefold in the past fifty years.

Without improvements in efficiency, growth of the demand for energy in these years would certainly have been larger. It is not clear how much larger the demand for energy would have been, since improvements in efficiency may provide incentive for growth by reducing the cost of some activities. (Thus, it would be a mistake to conclude that we would presently be using from two to five times as much energy in the absence of these improvements.) If future gains in efficiency do not continue as in the past, however, we should expect demand for energy to expand at a faster rate than a simple projection of the experience of the past fifty years would indicate.

It does not take an elaborate engineering argument to show that the improvements in efficiency of the past seventy years cannot be continued indefinitely. The present efficiency of energy utilization for the entire economy has been estimated to be about 50 percent.[2] This estimate is on the high side because it does not take account of losses in extracting, refining, transport, and distribution of fuel. Since we cannot get more energy out of an energy conversion process than we put in, a doubling of efficiency would carry us to the magical 100 percent level which occurs occasionally in textbooks but never in real engineering or industrial processes. The question, then, is what can we expect to achieve if a doubling of efficiency is an unreachable upper limit.

Futuristic dramas and science fiction take energy conversion problems rather lightly, and our spectacular achievements of the past should make anyone careful in pronouncing anything impossible.

Nevertheless, the limitations that we must live with for the next fifty or one hundred years seem already to be known. The theoretical efficiency of a heat engine is Efficiency = Initial Temperature - Final Temperature,
Initial Temperature
with temperature expressed in degrees Kelvin. (0° Kelvin = "absolute zero" = −460° F.) In order to approach 100 percent efficiency with such an engine it would be necessary to operate at initial temperatures near 2650° F and low temperatures near −453° F. These temperatures correspond with those of molten rock at the upper end and liquid helium at the lower. Such an engine, if it could be built, might have an efficiency above 99 percent, but it would not be of much use because more energy would be required to maintain the operating temperatures than could be obtained from running it.

Operating a heat engine between the freezing and boiling points of water yields efficiencies of only 35 to 40 percent. Theoretical efficiencies of about 60 percent might be attainable with a lower temperature of 100° F and an upper temperature of 1000° F. This is about the largest temperature range used in present-day electric power plants which, however, do not achieve 60 percent efficiencies because engineering problems of the real world eat into the theoretical maximum very rapidly. Since it is impossible to maintain a constant high temperature, the efficiency promptly drops below 55 percent. In the overall operation of the power plant, the efficiency of conversion from chemical fuel to heat (about 85 to 95 percent) must be included. And finally, the generators themselves — inherently highly efficient devices — are not 100 percent efficient. With current technology, efficiencies in electric power plants thus have a practical limit of 45 to 50 percent. The best operating plants do achieve efficiencies of 42 or 43 percent. We cannot go much further.

Thermionic devices and solar cells (see chapters 7 and 8) are occasional favorites of science fiction writers. But these too are heat engines and subject to the same theoretical limitations. Currently their efficiencies range between 10 and 15 percent. We will get no remarkable improvements from this source.

Owing to their use in the space program, fuel cells have received a good deal of attention. The separation of water into hydrogen and oxygen gases by solar energy might be a scheme of some interest for the future (chapter 7). The hydrogen and oxygen can be recombined to generate energy. Such a process might be attractive for transportation, for of presently realizable devices only the fuel cell promises a power source with a weight to horsepower ratio equal or superior to the ratio for the internal combustion engine. The hydrogen-oxygen fuel cell is also pollution-free, the only combustion product being water. Nevertheless, the theoretical maximum efficiency for such a fuel cell is about 80 percent and the best ones now available, which are far too expensive for routine use, achieve an efficiency of only 50 to 55 percent. It seems unlikely that this figure will improve very much. Such things as photosensitive semiconductors and thermoelectric devices may have their special uses, but their efficiencies also have rather stringent limitations.

Another way to look at gains in efficiency is to look at the principal uses of energy and inquire what gains might be made in specific areas. In transportation, the efficiency of the internal combustion engine has been remarkably resistant to improvement. Efficiencies of 21 or 22 percent characterized automobiles in the 1920s. Efficiencies today are about 25 percent, and any slight gains in utilization of fuel have been more than offset by the trend toward larger, heavier automobiles with overpowered engines. The hydrogen-oxygen fuel cell might produce a significant gain in engine efficiency, but that must be coupled with the efficiency of generating the fuel in the first place. While it is hard to estimate the efficiency of producing the fuel in the absence of a working scheme, one would expect the efficiency of the process to be less than 50 percent. In this case the overall efficiency of the fuel cell makes it a questionable improvement over the internal combustion engine, although it would represent a gain for pollution control.

We mentioned that the electric motor is very efficient, but as long as the efficiency of generating electricity at a central power plant is only 35 or 40 percent, the overall efficiency of electric automobiles will also be about the same as that of the present internal combustion engine. The switch to electric automobiles might be an attractive alternative, but not because of their efficiency.

Limits to efficiency are not the only limits nature imposes on us. The finiteness of our planet and its resources will also limit growth. Most of us think that what is familiar is normal and that what is normal must be right. Each one of us grew up and is living in times characterized by the runaway growth of most human institutions. This growth is so much a part of everyday life that we take it for granted, believing that growth must be desirable — even necessary. Many scientists, economists, and environmentalists, however, are now expressing concern over the tremendous growth in consumption of resources which has characterized the twentieth century. Growth in consumption is related to both population growth and increasing affluence. All forms of growth are coming under attack, for it is becoming more and more obvious that our planet cannot sustain unlimited growth.

Because growth is a matter of controversy and concern, it is important to understand some characteristics and results of the various kinds of growth. Growth usually occurs in one of three ways: at a constant rate, or at a rate which changes by a constant amount in each unit of time, or at a rate which changes with time by an amount which depends on the size of the growing thing. If you are given to saving dimes in a coffee can under your bed, three regimens of saving, representing the three modes of growth, would give the following results after one year:

Rate of Saving	Rate of Growth at End of Year	Total Savings at End of Year
10¢ per month	$ 0.10 per month	$ 1.20
Increase of 10¢ per month	1.20 per month	7.80
Double each month	204.80 per month	409.50

There are examples of each type of growth in everyday life. The first is illustrated by the way the bathtub fills up when you plug the drain and turn on the water. Or you may actually be squirreling away a dime a month under the bed. The second is illustrated by the acceleration of a falling object: the rate of descent increases 32 feet per second during each second of the fall, regardless of how fast the object is traveling or how long it has been on its way. The third type of growth is represented by the exponential growth rate that characterizes most living organisms and populations at one time or another, as well as many human institutions. The classic example is the compound interest offered by the savings and loan company.

It is helpful to think of exponential growth in terms of the length of time required to double the amount you started with. Below we list the doubling times at various rates of growth.

Annual Percent Increase	Doubling Time (years)
0.5	140
1.0	70
2.0	35
3.0	24
5.0	14
7.0	10
10.0	7

One thing that has been growing exponentially is the use of energy, both in the United States and in the world. It has been largely the growth of energy use that has permitted other kinds of growth — growth of food production, steel and aluminum production, transportation, communication, population, pollution, and general environmental degradation. Although some of the undesirable consequences of energy use may be unnecessary, they are realities, and as we look toward the future it seems probable that they will remain realities for some time.

The amount of energy used in the United States has doubled four times since the beginning of this century. It is expected to double twice again before the century's end. The rate of increase in energy use for the world is even greater, and can be expected to remain so, as people of poor nations strive for a decent life.

The results of exponential growth are always shocking. Many examples, some real, some facetious, dramatize what happens when growth proceeds unchecked. *Escherichia coli* is a common bacterium that, under appropriate conditions, reproduces itself by cell division every twenty minutes. Suppose that a student in microbiology inoculates one liter of nutrient medium with one milliliter of a culture of *E. coli* containing about one hundred living bacteria. From time to time, he withdraws a sample from the culture, makes a series of dilutions, and plates samples of these dilutions onto solid culture medium. The next day he counts the number of colonies that have grown on the plates, each colony representing the progeny of a single cell in the original culture.

After 3 ½ hours, the number of bacteria in the original culture had increased to around 3000; at 5 hours it was more than 50,000; at 6 ½ hours it was passing the million mark; at 8 hours, more than 25 million; and at 12 hours there were 100 billion cells! Then something happened. Samples taken during the next 24 hours continued to indicate that there were about 100 billion living cells. Then the number began to fall, slowly at first, then faster and faster, until another day went by and most of the cells were dead.

Then there was the clever lad who elected to receive, not a million dollars, but one cent the first day, two the second, four the third, eight the fourth, and so on every day for a month, until he was reckoning his wealth in millions of dollars. And the pair of grasshoppers that reproduced until their great great great . . . grandchildren formed a mass heavier than the earth. And finally, it has been pointed out that if science grows as it has been growing, in less than a hundred years there will be twice as many scientists as people.

Because projections of exponential growth curves always lead to absurdity, we tend not to take them seriously. We are alternately advised, "Don't worry, it won't happen," and "Do worry, it won't happen." The thing to remember is that the projection of the exponential phase of a growth curve is only a mathematical abstraction, not a prediction. Although it is certain that the absurdity will not come to pass, what happens instead should concern us. The bacteria ran out of food and poisoned themselves in their own wastes. Some of the grasshoppers brought the wrath of man and DDT upon themselves; most of the rest starved or were eaten by blackbirds. Only the little boy escaped disaster. Was it because he was a very intelligent boy, who stopped himself before natural limits stopped him?

There is urgency in the voices of those who warn of the immediacy of our growth-related problems. The reason for this urgency is incomprehensible to those who do not grasp the implications of exponential growth.

Human populations and institutions have been growing for hundreds of thousands of years. What, then, does this mean, that we have only a few more years to reverse the trends before nature enforces her own brutal controls? Imagine the germ of a rare disease — so rare that it infects only two-millionths (two ten-thousandths of a percent) of the earth's surface. Suppose that it doubles the area it infects each year and we arbitrarily decide to do something about it when it covers half the earth. After ten years of exponential growth it has spread over one one-thousandth of the earth, 0.1 percent. After fifteen years it is spreading faster, but still it covers only 3 percent of the earth. In the nineteenth year we are jolted to our senses: by the end of the year our germ will have reached the halfway point in its conquest, leaving us one year in which to save ourselves. What is so insidious about exponential growth is the shocking rapidity with which natural limits are reached after long periods of apparently modest and harmless increase. If you double your use of something each year, one year before it is all gone it is only half gone. Each year you use as much as you used in all the years before, combined.

There is no longer any doubt that the earth's resources are finite. Most disagreement among experts revolves around assigning numerical values to various limiting factors. But by now it should be clear that the difference between optimistic and pessimistic estimates of any resource becomes insignificant as long as we persist in growing. The predicament is illustrated by some simulation studies done at the Massachusetts Institute of Technology.[3] The question posed to the computer was, what will happen under various patterns of growth in world population, resource use, and pollution? The answer was clear. It makes little difference, in the long run, what assumptions are made. The day of reckoning for industrial society and for the world is hastened or delayed by only trivial amounts as long as we persist in any of the conventional types of growth. We are approaching a number of natural limits simultaneously and exponential growth will not continue much longer.

What will happen? This issue causes harsh controversy among scientists, economists, and political leaders, because a world without growth threatens many of our cherished institutions, beliefs, and theories. But inevitably we will deal with the problem, either by passive surrender to fate or by constructive action. Since the use of energy is fundamental to human culture, a first step in the control of growth might be the examination of our uses of energy and a voluntary restriction of energy use to the satisfaction of basic needs. Others may gaze into the cloudy crystal ball to foresee factors that might change growth patterns, either by design or by accident, and to foresee conditions that might accompany variously altered growth patterns. One thing is clear: the longer we postpone action to control growth, the fewer are the alternatives open to us. Increases in technical efficiency are almost exhausted, and technological innovation will mean mostly undirected social change if efficiency cannot be increased. The gains deriving from growth need a new balancing against the risks.

References

1. C. Summers, "The Conversion of Energy." *Scientific American* 224, 3 (1971): 148-160.
2. E. Cook, "The Flow of Energy in an Industrial Society," *Scientific American* 224, 3 (1971): 134-147.
3. D. H. Meadows, *et al.*, *The Limits to Growth* (New York: Universe Books, 1972).

4

THE ENERGY WE EAT

Times are changing, mister, don't you know? Can't make a living
on the land unless you've got two, five, ten thousand acres and a tractor.
Cropland isn't for little guys like us any more. You don't
kick up a howl because you can't make Fords, or because you're
not the telephone company. Well, crops are like that now. Nothing
to do about it.

John Steinbeck, 1939

Feed men, and then ask them of their virtue.

Fyodor Dostoevsky

The kind of energy that has always been of first importance is food.
Throughout most of history, man has relied on his own labor to provide
food. If the energy value of the food obtained had not substantially exceeded the energy expended in obtaining it, our species would not have
survived.

Through the development of agriculture, man was able to manipulate
the flow of energy in various ecosystems in order to divert an increasingly
large fraction of the earth's productivity to his own use. He did this by
simplifying the complex natural ecosystem — by decreasing the number of
species in the system and by controlling the species that competed with
him for the yield. Maintenance of the simplified system required an
endless input of energy. At first this was restricted to manpower, as
techniques for preparing the soil, planting, weeding, driving off pests, and

Giant combines move through a Washington wheat field. Farming with machinery is not new, but the scale and energy use of farm equipment continue to increase. (Photo courtesy of U.S. Department of Agriculture.)

harvesting were developed. Later, animal power was exploited. Still later, inanimate energy from wind and water was put to work, and finally, energy from the fossil fuels was utilized in food production.

In many parts of the world, agriculture still depends on energy from people, animals, and the sun. If conditions are favorable, such solar agriculture can give very high yields — for example, in wet rice culture, up to 50 calories are returned in harvest for every calorie of human energy investment. Modern agriculture, however, depends on converting fossil fuel into meat and potatoes, and there is an increasingly unfavorable ratio of energy input to food output. It is due to large-scale energy subsidy that agricultural yields have increased manyfold in the United States during this century. In 1900, a single farmer could feed about five people. By 1940 he could feed ten and in 1960 he could feed twenty-five. Today, the labor of one farmer feeds nearly fifty people, owing to the development of new fungicides, herbicides, rodenticides, insecticides, miticides, nematocides, antibiotics, vaccines, fertilizers, equipment, and specialized varieties of plants and animals. But it is not just one farmer who feeds fifty people. It is one farmer, many tons of coal, many barrels of oil, and an immense food processing and distribution system. In this chapter we will see how the present situation has come about.

Life can be viewed as a ceaseless web of energy conversions which follow well-known principles of energy conservation and transformation. Many ecologists describe an ecosystem in terms of the energy that flows into and out of it and from one trophic level to another (that is, from green plants to herbivores to carnivores or from any of these levels to decomposers). The tropical rain forest is an example of a complex ecosystem characterized by a large mass of living organisms and efficient use of available energy. In the rain forest there are many hundreds of species of plants, each specialized in its structure and function to fill a unique niche, and each accompanied by its cadre of insect predators specialized to deal with it. Armies of microorganisms live in association with roots, soil, and dead matter. Fungus-eating flies, beetles, termites, and other small animals feed on the microorganisms. They in turn are eaten by a variety of amphibians and reptiles. Birds and mammals feed on everything, including each other. Man, the most relentless and least specialized predator of all, is the one creature likely to upset the balance.

A great deal of energy is expended in maintaining the integrity of such a system. In the rain forest, energy flows through an intricate network of feedback loops that regulate all the interrelationships among species. Epidemics are virtually impossible because of the low density of any particular population. If one species increases, others usually increase to counteract it until order is restored. Similarly, should a species decline, its predators also decline or shift their attention to an alternative food source until decreased predation permits recovery of the original population. Even if one species should disappear, the system survives because there are many alternative pathways along which energy and materials can flow. Diversity of species insures that nutrients will be cycled effectively and soil structure preserved.

This complex, stable natural system supports the highest productivity and the greatest mass of living organisms possible under prevailing climatic conditions. Man can glean very little from this system, however, for little remains after respiration, predators, parasites, and decomposers claim their share. The situation in a rain forest is in sharp contrast to that in a field of grass. Energy is stored in a field of grass: it is used primarily for growth, rather than for maintaining the structure of the system. But a field of grass generally lacks stability because it contains relatively few species and few feedback loops which regulate interactions among species. Many ecological niches are empty, inviting other species to invade. Under many conditions, the short-lived, rapidly growing, prolific species in the open field tend eventually to be displaced by a succession of slower growing, longer-lived, less prolific species until a system develops which is relatively stable under the prevailing conditions of soil and climate. Left undisturbed, a meadow usually becomes a forest.

The achievement of agriculture is to simplify the system so that, as in a field of grass, the maximum amount of solar energy is channeled into growth. Man supplies the energy for maintenance. He plows and plants. He fights off competitors, predators, and disease. He supplies fertilizer to replace losses from harvesting. He shelters his animals, feeds and vaccinates them, and helps them to breed and to bear and raise their young. Thus, only through a steady input of energy is man able to divert most of the productivity of his fields to his own ends.

It is a difficult struggle to maintain the simplified system however much energy is put into it. Where fields lie barren for part of the year, erosion and deterioration of soil structure claim their toll. Bigger and better pests continually appear. Where irrigation is practiced, water shortages and increasing salinity haunt the farmer. And if a wheat field is abandoned, does a crop of wheat come up the next year? Of course not. This illustrates a major difference between wild plants and the plants that are adapted to modern agricultural methods. Wild plants are hardy, physiologically adaptable, and able to withstand adversity. Man's crops are highly specialized for growth, and consequently they are dependent on man for protection and even their own propagation.

The success of modern agriculture depends in part on the development of fast-growing, high-yielding strains of plants and animals, whose productivity in turn depends on energy-intensive agriculture. For more than ten thousand years man has engaged in selective breeding of domestic plants and animals. One of the most remarkable stories is that of maize, or corn. The first evidence of wild maize is from Mexico, dating from more than seven thousand years ago. The plant probably looked similar to any other grass that grows in meadows, and the ear was no larger than your thumbnail. Lurking in the small seed, however, was the potential to become the corn of today, with huge ears of closely packed grain. The yield of modern hybrid corn is more than 90 bushels per acre in the midwestern United States. But corn can no longer survive without man. Even if it weathered a dry season on poor soil, escaping disease and predation, it would perish within a season or two, for it can no longer shed its seed to propagate itself.

Other domestic plants and animals have also become increasingly dependent on man for their survival. It seems that an organism can be either a generalist or a specialist, but not both. An organism can process just so much energy in its lifetime. If this energy is channeled largely into growth, man must tend to the other needs of the organism. Our improved plants and animals are not truly improved, but are merely specialized. They are specialized in growing wool, fat roots, or giant fruits, in laying eggs or making milk. In diverting their energy into rapid growth, they sacrifice many qualities that allowed their wild forebears to survive. Selective breeding may solve specific problems, but we are fooling ourselves when we expect a plant to resist disease, discourage predators and competitors, withstand unfavorable climatic and nutritional conditions, grow rapidly, reproduce abundantly, and still yield a bountiful harvest that is tasty, nutritious, and beautiful. One manifestation of the penalty for improvement is that, in the plant world, favorite varieties of grapes, roses, and citrus, to name a few, are routinely grafted onto wild rootstocks to increase their vigor.

How hard the myth of the superplant or superanimal dies is shown by the case of the dog — the first animal domesticated. What has been done, in domesticating the wolf, is to segregate and intensify individual wolf traits in the various breeds of dog. There are dogs that surpass the wolf in size, strength, speed, or endurance, and in ability to track, guide, guard, retrieve, fight, swim, or jump, but no dog can do all these things better than a wolf. Some people dream of combining these skills into one superdog. Of course, the goal eludes them. What they are trying to do, in recombining all the talents that have been so carefully separated and amplified, is to reinvent the wolf.

Another feature of modern crop plants and domestic animals is genetic uniformity. Just as species diversity enhances a natural system's chances for survival, genetic diversity enhances a species' chances. If misfortune befalls a genetic subset of a group, the remaining members can avert disaster for the population as a whole. In addition to being specialized for growth of one sort or another, domestic plants and animals also tend toward genetic uniformity, which implies uniform resistance or susceptibility to adversity. Through modern practices in animal breeding, a dairy bull can spread his genes around the world before some camouflaged weakness makes itself known in his offspring. Vegetative propagation of plants and other common practices insure dependably uniform crops down to the end. For a popular variety of potato known as the lumper, the end was the Irish potato famine of 1846.

It is contradictory to desire uniformity and diversity at the same time. Modern agriculture values uniformity above all else. The risks are well known, but the benefits seem worth the gamble. As a result, there have been some disastrous epidemics in our country, including southern leaf blight of corn, fungus disease of a popular strain of oats called Victory, and the wildfire spread of a virulent new race of wheat stem rust which attacked a variety of wheat bred for resistance to the old race of wheat stem rust. The problem of plant breeders is that while they make hundreds of experimental crosses, nature tries millions. One mutant fungus spore or one fecund insect can undo years of the work of man.

The high-yielding varieties of grain that are being spread throughout the world as the "green revolution" represent the most extensive experiment in uniformity of crops yet tried. When things go well with the green revolution they go very well indeed; but the miracle grains are susceptible to widespread disaster and they produce no miracles unless grown under the most favorable conditions. But complex as are the problems of the green revolution, the alternative seems to be starvation for about a billion people.

There are three ways in which we can try to feed the growing number of human beings on earth: open up new agricultural and grazing lands, increase the productivity of the lands already in use, and develop new sources of food. The long-term success of any of these depends on the size of the population we are trying to support and on our ability to design self-sustaining systems for which there is a non-depletable source of energy and in which materials are recycled.

We are familiar enough with the surface of our planet to know that no hidden paradise remains to be discovered. The area of the earth's ice-free land surface is about 32 billion acres. Roughly one-quarter of this is potentially arable, another quarter is potentially grazable, and half is useless for agriculture, although some of it is forested. Much of the unused potentially arable land is in the tropics, where its potential may be more theoretical than practical. It certainly is not farmable with today's knowledge and techniques and economic restrictions. Practically all the land that can be cultivated under existing social and economic conditions is already under cultivation.

In North America, large amounts of arable land are not being farmed. However, we should not be optimistic about possibilities for expansion, for the best land is already cultivated and what remains has serious deficiencies. Furthermore, buildings and highways and other developments are spreading so malignantly over so much of our prime farmland that projections for California, for example, indicate that in less than fifty years half of the state's agricultural land will have been converted to nonagricultural uses. It is doubtful that, in opening new lands, we can run fast enough to stay where we are.

There are sound reasons for not rushing to open up vast new areas for crops or grazing, even if technical and economic obstacles could be overcome. One reason is that agriculture, especially as practiced in industrialized nations, depends on the activities of unmanaged ecosystems for the cycling of wastes and other materials. The capacities of some of these systems are already overtaxed, and further reduction in their size would intensify the problems. Another is that biologists emphasize the importance of maintaining reservoirs of wild plants and animals from which new and valuable domesticates may be developed. In general, the best plan seems to be to increase the productivity of lands already under cultivation, which inevitably requires an energy subsidy. This is the trend that agriculture has taken in the United States, although the main goal has been to maximize productivity per farmer rather than productivity per acre.

In a modern industrial society, only a tiny fraction of the population is in frequent contact with the soil, and an even smaller fraction of the population raises food on the soil. The proportion of the population engaged in farming halved between 1920 and 1950, and then halved again by 1962. Now it has almost halved again and yet a majority of the remaining farmers hold part-time jobs off the farm.[1] Simultaneously, work animals declined from a peak of more than 22 million in 1920 to very small numbers at present.[2]

In economic terms, the value of food as a portion of the total goods and services of society now amounts to a smaller fraction of the gross national product than it once did. Energy inputs to farming have increased enormously since 1920,[3] and the apparent decrease in farm labor is offset in part by the growth of support industries for the farmer. But with these changes on the farm have come a variety of other changes in the U.S. food system, many of which are now deeply embedded in the fabric of daily life. In the past fifty years, canned, frozen, and other processed foods have become principal items in our diet. At present the food processing industry is the fourth largest energy consumer in the Standard Industrial Classification groupings. The use of transportation in the food system has grown apace, and the proliferation of appliances in both numbers and complexity still continues in homes, institutions, and stores. Hardly anyone eats much food as it comes from the fields. Even farmers purchase most of their food from markets in town.

Present energy supply problems make this growth of energy use in the food system worth investigating. It is the purpose of this chapter to do so. But there are larger matters at stake. Georgescu-Roegen notes that "the evidence now before us — of a world which can produce automobiles, television sets, etc., at a greater speed than the increase in population, but is simultaneously menaced by mass starvation — is disturbing."[4] In the search for a solution to the world's food problems, the common attempt to transplant a small piece of a highly industrialized food system to the hungry nations of the world is plausible enough, but so far the outcome is unclear. Perhaps an examination of the energy flow in the U.S. food system as it has been developed can provide some insights that are not available from the usual economic measures.

Descriptions of agricultural systems are given most often in economic terms. A wealth of statistics is collected in the United States and in most other countries indicating production amounts, shipments, income, labor, expenses, and dollar flow in the agricultural sector of the economy. In what follows, we will make use of these statistics, for these values are of considerable use in determining the economic position of farmers in our society. But economic statistics are only a tiny fraction of the story.

Energy flow is another measure available to gauge societies and nations. Only after some nations shifted large portions of the population to manufacturing, specialized tasks, and mechanized food production, and shifted the prime sources of energy to fuels that were transportable and usable for a wide variety of activities could energy flow be a measure of a

Giant grain elevators, like these in Duluth, Minnesota, are only the first step in the storage, processing, and distribution portion of the food system. (Photo courtesy of Marine Studies Center, University of Wisconsin.)

society's activities. Today it is only in one-fifth of the world that these conditions are sufficiently advanced.

What we would like to know is: how does our food supply system compare in terms of energy use to other societies' and to our own past? Perhaps, knowing this, we can estimate the value of energy flow measures as an adjunct to, but different from, economic measures.

In the morning, breakfast offers orange juice from Florida by way of the Minute Maid factory, bacon from a midwestern meat packer, cereal from Nebraska and General Mills, eggs from California, milk from not too far away, and coffee from Colombia. All these things are available at the local supermarket (4.7 miles each way in a 300 horsepower automobile), stored in a refrigerator-freezer and cooked on an instant-on gas stove.

The present food system in the United States is complex and the attempt to analyze it in terms of energy use will introduce questions far more perplexing than would the same analysis performed on simpler societies. Such an analysis is worthwhile, however, if only to find out where we stand. We have a food system and most of us get enough to eat from it. If, in addition, one considers the food supply problems, present and future, of societies where a smaller fraction of the people get enough to eat, then our experience with an industrialized food system is even more important. There is simply no gainsaying the fact that most nations are trying to acquire industrialized food systems of their own, whether in whole or in part.

What economics tells us is that food in the United States is expensive by world standards. In 1970 the average annual per capita expenditure for food in the United States was about six hundred dollars.[5] This is larger than the per capita gross domestic product in more than thirty nations of the world. These thirty nations contain most of the world's people and a vast majority of those who are underfed. It would be convenient to know whether we can put our hands into the workings of our own industrialized food system to extract a piece of it that might mitigate their plight, or whether they must become equally industrialized in order to operate such a food supply system. Even if we consider the diet of a poor resident of India, the annual cost of his food at U.S. prices would be about two hundred dollars — more than twice his annual income.

The analysis of energy use in the food system begins with an omission. We will neglect that crucial input of energy provided by the sun to the plants upon which the entire food supply depends. Photosynthesis is about 1 percent efficient; thus the maximum solar radiation captured by plants is about 450 calories per square foot per year. Ultimately we can compare the solar input with the energy subsidy supplied by modern technology.

Seven categories of energy use on the farm were considered. The amounts of energy used are shown in figure 4-1. The values for farm machinery and tractors are for the manufacture of new units only and do

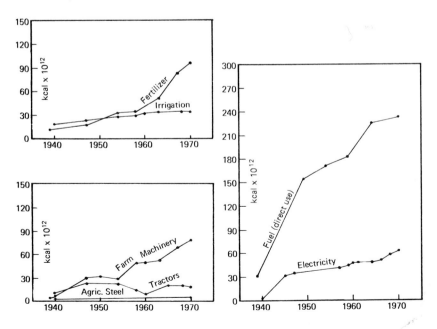

Figure 4-1. Energy use on farms, 1940-1970. Transportation is included in the food processing sector.

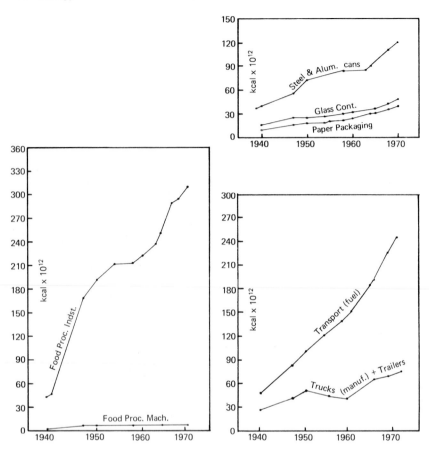

Figure 4-2. Energy use in the food processing sector.

not include parts and maintenance for existing units. The amounts shown for direct fuel use and electricity consumption are a bit too high because they include some residential uses of the farmer and his family. Note the relatively high energy cost associated with irrigation. In the United States, less than 5 percent of the cropland is irrigated. In some countries where the green revolution is being attempted, the new high yield varieties require irrigation while native crops did not. If that were the case in the States, irrigation would be the largest single use of energy on the farm.

Little food makes its way from field and farm directly to the table. The vast complex of processing, packaging, and transport has been grouped together in a second major subdivision of the food system. Figure 4-2 displays the energy use in food processing and packaging. Energy use for transport of food should be charged to the farm in part, but we have not done so because the calculation of the energy values is easiest (and we believe most accurate) if they are taken for the whole system.

After food is processed there is further energy expenditure. Transportation enters again, and some fraction of the energy used for transportation should be assigned here. But there are also the distributors, wholesalers and retailers, whose freezers, refrigerators and very establishments are an integral part of the food system. There are also the restaurants, schools, universities, prisons, and a host of other institutions engaged in the procurement, preparation, storage, and supply of food. We have chosen to examine only three categories: the energy used for home refrigeration and cooking, for commercial refrigeration and cooking, and that used for the manufacture of the refrigeration equipment. Figure 4-3 shows energy consumption for these categories. There is no attempt to include the energy

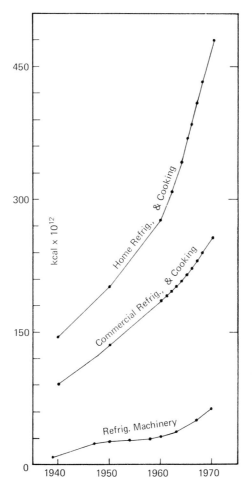

Figure 4-3.   Commercial and home energy use in the food system. These are selected uses only.

used in trips to the store or restaurant. Garbage disposal has also been omitted, although it is a persistent and growing feature of our food system. Twelve percent of the nation's trucks are engaged in waste disposal which is largely, though not entirely, related to food. If there is any lingering doubt that these activities — both the ones included and the ones omitted — are an essential feature of our present food system, one need only ask

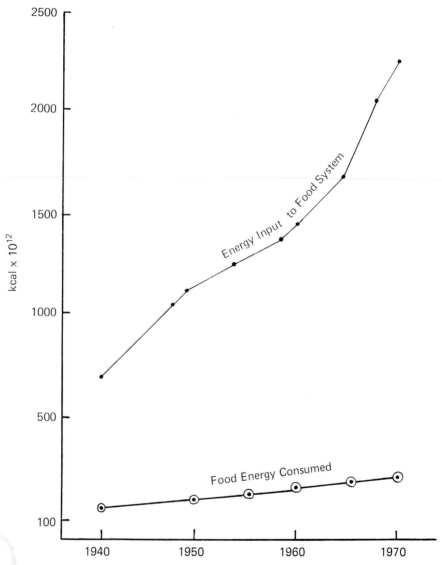

Figure 4-4. Energy use in the food system, 1940-1970, compared to caloric energy content of food consumed.

what would happen if everyone should attempt to get on without a refrigerator or freezer or stove? Certainly the food system would change.

We have summarized the numerical values for primary energy used by the U.S. food system, from 1940 to 1970. As for many activities in the past few decades, the story is one of continuing increase. The totals are displayed in figure 4-4 along with the energy value of the food consumed by the public. The food values were obtained by multiplying the annual caloric intake with the population. The difference in caloric intake over this thirty-year period is not significant and the curve mostly indicates population increase.

The difficulty with history as a guide for the future or even the present lies not so much in the fact that conditions change — we are at least continually reminded of that fact — but that history is only one experience of the many that might have been. The U.S. food system developed as it did for a variety of reasons, many of them probably not understood. It would be well to examine some of the dimensions of this development before attempting to theorize about how it might have been different, or how parts of this food system can be transplanted elsewhere.

Figure 4-5 displays features of our food system not easily seen from economic data. The curve shown has no theoretical basis, but is suggested by the data as a smoothed recounting of the history of increasing food production. It is, however, similar to growth curves of the most general

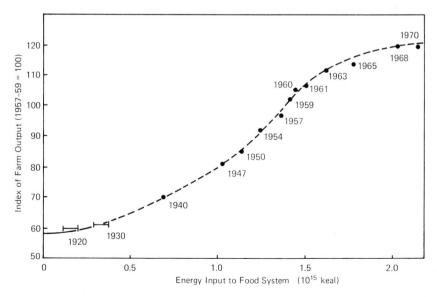

Figure 4-5. Farm output as a function of energy input to the U.S. food system, 1920-1970.

kind, and it suggests that, to the extent that the increasing energy sub-sidies to farm production have increased that production, we are near the end of an era. This growth curve demonstrates an exponential phase which began in 1920 or earlier and lasted until 1950 or 1955. Since then the in-crements in production obtained by the growth in energy use have become smaller. It is likely that further increases in food production from in-creasing energy inputs will be harder and harder to come by. Of course, a major modification in the food system could change things. However, the argument advanced by the technological optimists — that we can always produce more if we have enough energy and that no other major changes are needed — is not supported by our own history.

One farmer now feeds fifty people, and the common expectation is that labor inputs to farming will continue to decrease in the future. Behind this expectation is the assumption that continued application of technology — and energy — to farming will substitute for labor. Figure 4-6 is the substitution curve of energy for labor on the farm. It shows the

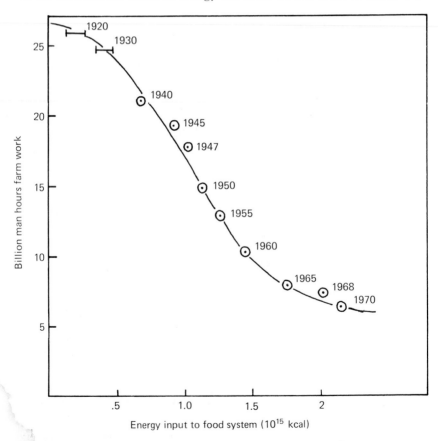

Figure 4-6. Labor use on farms as a function of energy use in the food system.

This, too, is part of the U.S. food system. As labor on the farm has dropped off, new food-related jobs have expanded. Yesterday's farm youth is today's fast food carhop. (Photo by Sandy Levitz.)

historic decline in farm labor as a function of the energy subsidy to the food system. Again the familiar "S" shaped curve may be seen. Reduction of farm labor by increasing energy inputs cannot go much further.

The food system that has grown during this period has provided a great deal of employment that did not exist twenty, thirty, or forty years ago. Perhaps even the idea of a reduction of labor input is a myth when the food system is viewed as a whole, instead of examining the farm worker only. Pimentel and associates cite an estimate of two farm support workers for each person employed on the farm.[6] To this must be added employment in food processing industries, in food wholesale and retail establishments, and in the manufacturing enterprises that support the food system. Yesterday's farmer is today's canner, tractor mechanic, and fast food carhop. The process of change has been painful to many ordinary people. The rural poor, who could not quite compete in the industrialization of farming, migrated to the cities. Eventually they found other employment, but one must ask if the change was worthwhile. The answer to that question cannot be provided by energy analysis any more than by economic data, because it raises fundamental questions about how individuals would prefer to spend their lives. But if there is a stark choice of long hours as a farmer or shorter hours on the assembly line of a meat packing plant, it seems clear that the choice would not be universally in favor of the meat packing plant. Thomas Jefferson dreamed of a nation of

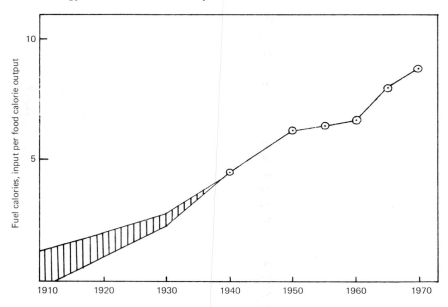

Figure 4-7. Energy subsidy to the food system to obtain one food calorie. Since the 1910-1937 values cannot be fully documented, we have presented a range for that period.

independent small farmers. It was a good dream, but society did not develop in that way. Nor can we turn back the clock to recover his dream. But in planning our future, we had better look honestly at our collective history, and then each of us look closely at his own dreams.

The data on figure 4-5 can be combined to show the energy subsidy provided to the food system for the recent past. We take as a measure of the food supplied, the caloric content of the food actually consumed. This is not the only measure of the food supplied, as many protein-poor peoples of the world clearly show. Nevertheless the ratio of caloric input to output is a convenient way to compare our present situation with the past. Figure 4-7 shows the history of the U.S. food system in terms of the number of calories of energy supplied to produce one calorie of food for actual consumption. It is interesting and possibly frightening to note that there is no indication that this curve is leveling off. Fragmentary data for 1972 suggest that the increase continued unabated. We appear to be increasing the energy input even more. Note that a graph like figure 4-7 could go to zero. A natural ecosystem has no fuel input at all, and those primitive people who live by hunting and gathering have only the energy of their own work to count as input.

The markets for farm commodities in the United States come closer than most to the economist's ideal of a free market. In a free market there are many small sellers and many buyers, and thus no individual is able to affect the price by his own actions in the marketplace. But a market would satisfy these conditions only in the absence of intervention in its function. Government intervention in the prices of agricultural products (and hence of food) has been a prominent feature of the U.S. food system for at least thirty years. Between 1940 and 1970 total farm income has ranged from 4.5 to 16.5 billion dollars, and that part of the National Income having its origin in agriculture (which includes indirect income from agriculture) has ranged from 14.5 to 22.5 billion dollars. Meanwhile government subsidy programs, primarily farm price supports and soil bank payments, have grown from 1.5 billion in 1940 to 6.2 billion dollars in 1970. In 1972 these subsidy programs had grown to 7.3 billion dollars despite foreign demand for U.S. agricultural products. Viewed in a slightly different way, direct government subsidies have accounted for 30 to 40 percent of farm income and they have accounted for 15 to 30 percent of the National Income attributable to agriculture for the years since 1955. The point is important because it emphasizes once again the striking gap between the economic description of society and the economic models used to account for that society's behavior.

The issue of farm price supports is related to energy in two ways. First, government intervention in the food system is a feature of almost all highly industrialized countries (although farm incomes lag behind national averages despite the intervention). Second, reduction of the energy subsidy to agriculture (even if we could manage it) might further reduce farmers' incomes. One reason for this state of affairs is that quantitative demand for food has definite limits and, without farm price supports, the only way to increase farm income is to increase the unit cost of agricultural products. Consumer boycotts and protests in the early 1970s suggest that there is considerable resistance to this course of action. Thus, government intervention in the functioning of the market for agricultural products has increased with the use of energy in agriculture and the food supply system and we have nothing but theoretical suppositions to suggest that either event could happen alone.

We have tried to analyze the complex, industrialized food system of the United States because of its implications for future energy use. But the world is short of food, and we must also consider the role of energy in feeding a growing and undernourished world population. A few years ago it was widely predicted that the world would experience widespread famine in the 1970s. The adoption of new high-yield varieties of rice, wheat, and other grains has caused some experts to predict that the threat of these expected famines can now be averted — perhaps indefinitely. Yet, despite increases in grain production in some areas, the world still seems to be headed towards famine. The adoption of these new varieties of grain — dubbed hopefully the green revolution — is an attempt to export a part of the energy-intensive food system of the highly industrialized countries to non-industrialized countries. It is an experiment, because the whole food system is not being transplanted to new areas, but only a small part of it. The green revolution requires a great deal of energy. Many of the new grain varieties require irrigation in places where traditional crops did not, and almost all the new crops require extensive fertilization. Both irrigation and fertilization require high inputs of energy.

The agricultural surpluses of the 1950s have largely disappeared. Grain shortages in China and the U.S.S.R. have attracted attention because they have brought foreign trade across ideological barriers. There are other countries that would probably import considerable grain if they could afford it. But only four countries may be expected to have any substantial excess agricultural production: Canada, New Zealand, Australia, and the United States. None of these is in a position to give grain away, because they need the foreign trade to avert ruinous balance of payment deficits. Can we then export energy-intensive agricultural methods instead?

It is quite clear that the United States food system cannot be exported intact at present. For example, India has a population of 550 million persons. To feed the people of India at the United States level of about 3,000 kilocalories per day (instead of their present 2,000) would require more energy than India now uses for all purposes. If we wished to feed the entire world with a food system of the U.S. type, almost 80 percent of the world's annual energy expenditure would be required.

The recourse most often suggested is to export only methods of increasing crop yield, and to hope for the best. We must repeat that this is an experiment. We know that our food system works (albeit with some difficulties and warnings for the future) but we do not know what will happen if we take a piece of that system and transplant it to a poor country that is lacking the industrial base of supply, transport system, processing industry, appliances for home storage and preparation, and most of all, a level of industrialization permitting higher food costs.

The energy requirements of green revolution agriculture have some important political and social implications. To the extent that the Western, highly industrialized countries must continue research and development for the new strains continually required to respond to new plant diseases and pests that can and do sweep through areas planted with a single variety (consider the recent problem with corn blight in the midwest), the Western countries will possess a hold over the developing countries. Political radicals sometimes dub this state of affairs "technological imperialism," but, whatever the name, the developing countries resent their dependence upon the vagaries of another nation's priorities. In order to avoid this source of friction the improved agriculture must be managed within the developing countries. In many of the developing countries such internal programs have begun. But establishment of anything like the agricultural extension network of the United States will require a significant expenditure of energy. Failure to establish networks of this type has, in some green revolution areas, favored the better-educated farmer against the peasants, who have little access to or knowledge of the new grain varieties. The necessity to fertilize and irrigate also favors the larger, more affluent farms — often with the result of driving more peasants off the land and into the cities, where developing nations face a difficult problem already.

Fertilizers, herbicides, pesticides, and in many cases, machinery and irrigation are needed to give any hope of success to the green revolution. Where is the energy for this to come from? Many of the nations with the most serious food problems are also those with scant supplies of fossil fuels. In the industrialized nations, solutions to the energy supply problems are being sought in nuclear energy. This technology-intensive solution, even if successful in advanced countries, poses additional problems for underdeveloped nations. To create the base of industry and technologically sophisticated people within their own country will be beyond many of them. Once again they face the prospect of depending upon the good will and policies of industrialized nations. Since the alternative could be famine, their choices are not pleasant, and their irritation at their benefactors — ourselves among them — could grow to threatening proportions. It would be comfortable to rely on our own good intentions, but our good intentions have often been unresponsive to the needs of others. The matter cannot be glossed over lightly. World peace may depend upon the outcome.

Application of energy on our farms is now near 100 kilocalories per square foot per year for corn,[7] and this is more or less typical of intensive agriculture in the United States. With this application of energy we have achieved yields of 200 kilocalories per square foot per year of usable grain — bringing us to almost half of the photosynthetic limit of production. Further applications of energy are likely to yield little or no increase in the level of productivity. In any case research is not likely to improve the efficiency of the photosynthetic process itself. There is a further limitation on improvement of yield. Faith in technology and research has at times blinded us to the basic limitations of the plant and animal material with which we work. We have been able to emphasize desirable features already present in the gene pool, and to suppress others that we find undesirable. At times the cost of increased yield is the loss of desirable characteristics — hardiness, resistance to disease and adverse weather and the like. The further we get from characteristics of the original plant and animal strains, the more care — and energy — is required. Choices must be made in the directions of plant breeding. And the limitations of the plants and animals we use must be kept in mind. We have not been able to alter the photosynthetic process, or to change the gestation period of animals. In order to amplify or change an existing characteristic we will probably have to sacrifice something in the overall performance of the plant or animal. If the change requires more energy, we could end with a solution that is too expensive for people who need it most. These concerns are intensified by the degree to which energy becomes more expensive in the world market.

Figure 4-8 shows the energy subsidy ratio to energy output for a number of widely used foods in a variety of times and cultures. For comparison the overall behavior of the United States food system is shown, but the comparison is only approximate because, for most of the specific crops, the energy input ends at the farm. As has been pointed out, it is a long way from the farm to the table in industrialized societies. Several things are immediately apparent, and coincide with expectations. High protein foods, such as milk, eggs, and meat, have a far poorer energy return than do plant foods. Because protein is essential for human diets, and because the amino acid balance necessary for good nutrition is not found in most cereal grains, we cannot abandon meat sources altogether. Figure 4-8 indicates how unlikely it is that increased fishing or production of fish protein concentrate will solve the world's food problems. Even if we leave aside the question of whether the fish are available — a point on which expert opinions differ — it would be hard to imagine, with rising energy prices, that fish protein concentrate will be anything more than a by-product of the fishing industry, for it requires more than twice the energy of production of grass-fed beef or eggs. Distant fishing is still less likely to solve food problems. On the other hand, near coastal fishing is relatively low in energy cost. Unfortunately, however, coastal fisheries are threatened with overfishing as well as pollution.

The position of soybeans may be crucial in figure 4-8. Soybeans possess the best amino acid balance and protein content of any widely grown crop. This has long been known to the Japanese, who have made soybeans a staple of their diet, and to beef feedlot operators. Are there other plants, possibly better suited for local climates, which have adequate proportions of amino acids in their proteins? There are about eighty thousand edible species of plants, of which only about fifty are actively cultivated on a large scale (and 90 percent of the world's crops come from only twelve species). We may yet be able to find species that can help the world's food supply.

The message of figure 4-8 is simple. In primitive cultures, 5 to 50 calories of food were obtained for each calorie invested. Some highly civilized cultures have done as well and occasionally better. In sharp contrast, industrialized food production requires an input of 5 to 10 calories of fuel to obtain 1 calorie of food. We must pay attention to this difference — especially if energy costs increase. If some of the energy subsidy for food production could be supplied on-site, with renewable sources — primarily

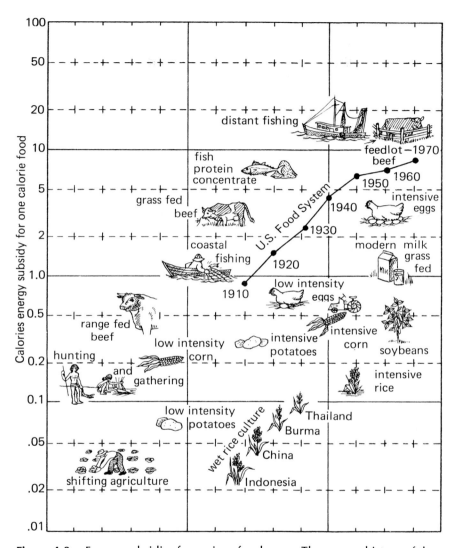

Figure 4-8. Energy subsidies for various food crops. The energy history of the U.S. food system is shown for comparison.

sun and wind — we might be able to provide an energy subsidy. Otherwise the choices appear to be less energy-intensive food production or famine for many areas of the world.

It is possible to reduce the amount of energy required for agriculture and the food system. A series of thoughtful proposals by Pimentel and associates deserves wide attention.[8] Many of these proposals mitigate environmental problems, and any reductions in energy use directly reduce

Feedlot beef is among the most energy-intensive foods. This large feedlot near Amarillo, Texas, can accommodate 70,000 head of cattle. A return to grassfed beef would use less energy and eliminate the manure disposal problem. With higher energy costs it might also be cheaper. (Photo courtesy of U.S. Department of Agriculture.)

the pollution due to fuel consumption while giving us more time to solve our energy supply problems. Among the suggestions made by Pimentel and associates are the following.

First, we should use natural manures. The United States has a pollution problem due to runoff from animal feedlots, and yet we apply large amounts of manufactured fertilizer to fields. More than one million kilocalories per acre could be saved by substituting manure for manufactured fertilizer (and as a side benefit, the condition of the soil would be improved). Widespread use of natural manure will require decentralization of feedlot operations so that manure is generated closer to the point of application. Decentralization would increase feedlot costs, but if energy prices rise, feedlot operations will rapidly become more expensive in any case. Crop rotation is less widely practiced than it was even twenty years ago. Increased use of crop rotation or interplanting winter cover crops of legumes (which fix nitrogen as a green manure) saves 1.5 million kilocalories per acre compared to commercial fertilizer.

Second, weed and pest control could be accomplished at a much smaller cost in energy. A 10 percent saving of energy in weed control could be obtained by using the rotary hoe twice in cultivation instead of using herbicides (again with pollution abatement as a side benefit). Biologic pest control — that is, the use of sterile males, introduced predators and the like — requires only a tiny fraction of the energy needed for pesticide manufacture and application. A change to a policy of "treat when and where necessary" in pesticide application would bring a 35 to 50 percent reduction in pesticide use. Hand application of pesticides requires more labor than machine or aircraft application, but the reduction of energy is from 18,000 kilocalories per acre to 300 kilocalories per acre. Changed cosmetic standards, which in no way affect the taste or edibility of foodstuffs, could also bring about a substantial reduction in pesticide use.

Third, the directions in plant breeding might emphasize hardiness, disease and pest resistance, reduced moisture content (to end the wasteful use of natural gas in drying crops), reduced water requirements, and increased protein content — even if it means some reduction in overall yield. In the longer run, plants not now widely cultivated might receive some serious attention and breeding efforts.

The direct use of solar energy on farms, a return to wind power (using the modern windmills now in use in Australia), and the production of methane from manure are all possibilities. These methods require some engineering to be economically attractive, but it should be emphasized that these technologies are now better understood than is the technology of breeder reactors. If energy prices continue to rise, these methods of energy generation would be attractive alternatives even at present costs of implementation.

Beyond the farm, but still far short of the table, many more energy savings could be introduced. The most effective way to reduce the large energy requirements of food processing would be a change in eating habits towards less highly processed foods. The current dissatisfaction with many processed foods from "marshmallow" bread to hydrogenated peanut butter could presage such a change, if it is more than just a fad. Technological changes could reduce energy consumption on an industry by industry basis, but the most effective way to encourage the adoption of methods using less energy would be to increase the cost of energy. Continuing price increases almost certainly await us.

Packaging has long since passed the stage of simply holding a convenient amount of food together and providing it with some minimal protection. Legislative controls may be needed to end the spiralling competition of manufacturers in amount and expense of packaging. In any case, recycling of metal containers and wider use of returnable bottles could reduce this large energy use.

The trend toward the use of trucks in food transport to the virtual exclusion of trains should be reversed. By simply reducing the direct and indirect subsidies to trucks, we might go a long way toward enabling trains to compete.

Finally, in the home we may have to ask whether the ever larger frostless refrigerators are needed, and whether the host of kitchen appliances really means less work.

Store delivery routes, even by truck, would require only a fraction of the energy used by private autos for food shopping. Rapid transit, giving some attention to the problems of shoppers with parcels, would be even more energy-efficient.

If we insist on a high-energy food system, we should start with coal, oil, garbage — or any other source of hydrocarbons — and produce food in factories from bacteria, fungi, and yeasts. These products could be flavored and colored appropriately for cultural tastes. Such a system would be more efficient in use of energy, solve waste problems, and permit much or all of the agricultural land to be returned to its natural state.

If energy prices rise — as they have already begun to do — the rise in the price of food in societies with industrialized agriculture can be expected to be even larger than the energy price increases. Slesser, in examining the case for England, suggests that a quadrupling of energy prices in the next forty years would bring about a sixfold increase in food prices.[9] Even small increases in energy costs may make it profitable to increase labor input to food production. Such a reversal of the fifty-year trend toward energy-intensive agriculture would present environmental benefits as a bonus.

We have tried to show how analysis of the energy flow in the food system illustrates features of the food system that are not easily deduced from the usual economic analysis. Despite some suggestions for lower intensity food supply and some frankly speculative suggestions, it would be hard to end this chapter on a note of optimism. The world drawdown in grain stocks which began in the mid-1960s continues, and some food shortages are likely throughout the 1970s and early 1980s. Even if population control measures begin to limit world population, the rising tide of hungry people will be with us for some time.

Food is basically a net product of an ecosystem, however simplified. Food starts with a natural material, however modified later. Injections of energy (and even brains) will carry us only so far. If mankind cannot adjust its wants to the world in which it lives, there is little hope of solving the food problem. In that case the problem will solve mankind.

REFERENCES

1. Statistical Abstract of the United States, 1973.
2. Historical Statistics of the United States (Washington, D.C., 1960).
3. D. Pimentel et al., "Food Production and the Energy Crisis," Science 182 (1973): 443-449.
4. N. Georgescu-Roegen, The Entrophy Law and the Economic Process (Cambridge, Mass.: Harvard University Press, 1971), p. 301.
5. Pimentel, op. cit.
6. Ibid.
7. Ibid.
8. Ibid.
9. M. Slesser, "How Many Can We Feed?," The Ecologist 3 (1973): 216-220.

PART TWO

ENERGY RESOURCES
AND THEIR USE

5

THE FOSSIL FUELS

The time has come to inquire seriously what will happen when our forests are gone, when the coal, the oil, and the gas are exhausted.

Theodore Roosevelt, 1908

Industrial civilization depends on the continuous availability of work and power supplied by machines, which in turn require a constant supply of energy. Since the invention of the steam engine, most of this energy has been obtained by the burning of fuel. Energy and fuel are so nearly synonymous in our thinking that we are dismayed by the prospect of switching to new types of energy for industrial and domestic use. We cannot take for granted a comfortable future based on nuclear fission, nuclear fusion, or solar energy because we have not yet mastered these technologies; and unless we master them before the fires of our fuel-based culture die out we will never do it.

When the history of man is complete at some unknown date in the unknowable future, the period in which he burned the fossil fuels will be a brief chapter. After two million years on this planet, man discovered first "sea coales," then oil shale (the "rocks that burn") and petroleum. Then, in an unbridled spree, he dug up the earth wide and deep and, within fifteen hundred years, burned all that he could extract. The place of the fossil fuels in man's cultural history is illustrated in figure 5-1.

The fossil fuels include the petroleum family of hydrocarbons classified as crude oil, natural gas, and tar sands, and also coal, lignite, and a related material known as oil shale. We will look at these individually, considering the changing pattern of relationships among them.

Half a billion years ago, when the continents were still bleak and lifeless expanses of rock, the oceans already teemed with life. Plants were simple, although some of them reached great size, for their watery home did not evoke the range of chemical and structural adaptations that future land plants would devise. Representatives of all major groups of animals except the vertebrates were present; but even then, there was something that would eventually become a fish. Marine life, then as now, was most abundant in shallow seas and coastal waters, where light could penetrate to support photosynthesis and thus provide food for non-photosynthetic creatures.

Half a billion years ago, anaerobic microorganisms inhabited the stagnant ooze that collected in hollows on the sea floor. These organisms were probably evolution's refugees from a prephotosynthetic past, when free oxygen was absent from the earth. They needed oxygen to maintain their metabolic fires and they got it by extracting it from organic molecules.

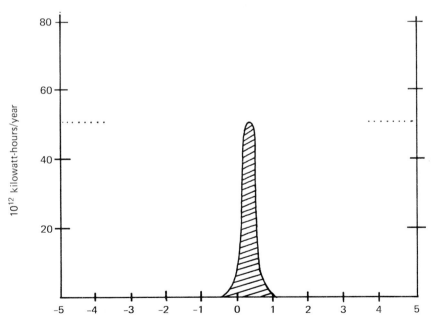

Time before and after the present, in thousands of years

Figure 5-1. The period of exploitation of the fossil fuels, in the time frame of human history. (Redrawn by permission from M. K. Hubbert, "Energy Resources," in *Resources and Man*, National Academy of Sciences-National Research Council [San Francisco: W. H. Freeman and Co., 1969], p. 206.)

Plants and animals lived and died in oxygenated waters and their corpses settled into the bottom mud, where anaerobic decomposers fed on them. A residue of hydrocarbon molecules (compounds of hydrogen and carbon) was left behind. The process was slow and energetically inefficient, but time is plentiful in earth history. Continuing accumulation of sediments increased the temperature and pressure in layers below, causing further molecular rearrangements to give a medley of straight-chain, branched-chain, and ring-shaped molecules. Thus time, microbial action, heat, and pressure changed cast-off parts of organic molecules to petroleum — one of the most complex natural mixtures to be found on earth.

Although the oldest known petroleum was formed at least half a billion years ago, most of the commercially valuable deposits are less than half that old. Early stages of petroleum formation are still in progress in the Black Sea and the Gulf of Mexico, where studies of bottom sediments show that the first recognizable stages of conversion of organic sediments into oil require' three to nine thousand years.

The sediments where petroleum is formed contain a high proportion of sea water, in which droplets of petroleum are dispersed. With time, these sediments are buried under further layers of mud and organic matter. As the rocks become consolidated, the liquid migrates outward and upward in response to movements of the earth or variations in pressure from above. Eventually, the petroleum separates from the water, either rising higher by virtue of its lower density, or being held back because of its greater viscosity. Thus petroleum leaves the rocks in which it was formed and moves through porous formations until its further progress is blocked by a roof of impervious rock. There it collects in "oil pools," which are not pools at all but rocks whose pores are filled with oil instead of groundwater.

Constituents of petroleum range from tarry and waxy compounds with molecular weights of 300 or more and boiling points well above 900° F to methane with a molecular weight of 16. They also include variable amounts of molecules containing oxygen, nitrogen, and sulfur.

The lighter molecules like methane are gases under usual conditions of temperature and pressure. If more gas is present than the liquid petroleum can hold in solution, it bubbles to the top and forms a gas cap over the oil pool. The gas, oil, and underlying groundwater are often under high pressure. This pressure is one cause of oil well gushers or blowouts, which were romanticized in the past by Hollywood but which today spell tragedy. In a properly controlled well, the natural pressure aids in the removal of oil or gas. Today, drilling and production of an oil field are carefully planned to take advantage of the reservoir pressure.

Wherever a crack in the impermeable cover of an oil or gas reservoir provides a passage for escape, oil or gas will escape. Furthermore, if, through erosion or another process, covering rocks are stripped away, the pressurized petroleum may force its own passageway through the roof of rock. Natural seeps of oil and gas and surface deposits of asphalt (the tarry residue left behind after volatile constituents evaporate) have been known since prehistoric times. Along the coast of the Caspian Sea, blazing jets of escaping natural gas were worshipped by ancient fire worshippers. Perhaps the fiery furnace into which King Nebuchadnezzer was wont to toss people was a burning seepage of natural gas, for the "Eternal Fires" of Mesopotamia were a curiosity in biblical times as they are today. The Egyptians used material from natural petroleum seeps for the preparation of mummies; the Greeks used it for weapons. More than five thousand years ago, the Sumerians drove their chariots over roads paved with asphalt. In the Middle East and on the Pacific coast of America, ancient peoples used asphalt to caulk their boats. And although the 69-foot hole dug by Colonel Edwin Drake in 1859 is heralded as the first oil well, the Burmese had been collecting oil from surface exudations and shallow hand-dug wells for centuries. So although widespread use of petroleum as a fuel dates back little more than one hundred years, petroleum products have been important in human culture since antiquity. Our recent habit of burning this resource is considered by some people as squandering a precious material that has many more vital uses.

As layers of rock formed while the continents rose from the seas only to be submerged, rise, and be submerged again, and as areas of the solid earth cracked, buckled, and slid across one another, potential oil and gas traps became arranged in stacks. Eventually oil migrated to these traps, each pool constrained beneath its impenetrable roof of caprock. The story of oil production is largely the story of success in tapping deeper pools by drilling to greater and greater depths, from that first well of 69 feet to modern wells that pierce the earth to a depth of 5 miles or more.

While basic drilling technology has changed relatively little, discovery of oil and gas fields has become an increasingly complicated venture, requiring the skills of geologists and geophysicists as well as engineers. The first oil fields were discovered by chance or by digging in the vicinity of natural seeps. The modern oil industry began in 1848, when drillers searching for water struck oil instead. Then Drake decided to drill for oil on purpose. After much difficulty in mustering and maintaining support for his project, he finally achieved his goal. Soon, "wildcatters" were sinking wells at sites chosen merely on a hunch. A few of these paid off fabulously, but most yielded nothing. As early as 1855, however, geologists were pointing out the tendency of oil pools to occur in certain types of geological structures. The rational search for oil then became the search for potential oil traps. Today, teams of earth scientists employ geologic, seismic, magnetic, gravitational, and electrical techniques to dog

In earlier days, oil fields were forests of oil derricks almost covering the surface of the ground. Shown here is the Spring Hill oil field in 1926. (Photo
courtesy of Shell Oil Co.)

oil into its last hideouts: the bottom of the sea, the hostile desert, and the
forbidding subpolar regions.

The quest for oil was originally sparked by a growing scarcity of whale
oil and other illuminants. It was the kerosene fraction that was in demand,
although there was a small market for lubricating oils and oil for medicinal
purposes. The infant petroleum industry mounted a sustained effort to
find and create new markets for other refinery products. The story of
petroleum in nineteenth century America is a story of vast overproduction,
profligate waste, falling prices, and a frantic struggle to stimulate new uses
for a product that could hardly be given away. With the advent of electric
lighting, the situation for petroleum would have been grim indeed had it
not been for three circumstances: the demonstration that oil could fire a

A modern oil field near Kingfisher, Oklahoma. Wider spacing of wells not only requires less land, but ensures maximum recovery of petroleum. (Photo by C. E. Rotkin — Texaco, Inc.)

boiler as well as coal, the shift of oil production to California, Texas, and Oklahoma, where coal was scarce and expensive, and the rise of the automobile.

By 1910, fuel oil was the number one refinery product and the production of kerosene was falling rapidly. By the end of World War I, fuel oil had made heavy inroads into the coal market in manufacturing, shipping, and railroads. Then came a rapid expansion of oil for domestic use and for the generation of electricity. During World War I, the demand for gasoline began to increase even more rapidly than the demand for fuel oil, as Americans began their love affair with the internal combustion engine. In the late 1920s gasoline gained first place among refinery products and it has been there ever since.

Production of crude oil in the United States and in the world is growing exponentially. Although we know that exponential growth cannot continue indefinitely, planners tend to confuse the projection of a growth curve with a prediction, and the prediction becomes a self-fulfilling prophecy. Predictions of the demand for oil for 1980 and 2000 are running headlong into predictions of chronic shortages.

There is confusion about the magnitude of our oil reserves, stemming from misunderstanding of the word reserves. Proved reserves of oil (or of any depletable resource) are the amounts proved by exploration to exist and known to be recoverable by methods that are technically and economically feasible. For economic reasons, the rate of oil exploration is geared to the rate of production. So when we are told that our proved reserves will last for twenty years at current rates of consumption, this does not mean that we will run out of oil in twenty years. Because the word reserves has been interpreted as "the amount we have left," however, we have become complacent because new discoveries always come along to shove the day of reckoning well into the future. The ceiling we face is not the current value of proved reserves but the total (unknown but finite) amount of recoverable oil that exists.

While others are squabbling about numbers and the significance of new discoveries and technologies, let us take an analytical look at trends in the oil industry. In the forty-eight contiguous states of the United States, we may already have reached the peak of oil production. Although the untapped resources of Alaska might postpone the peak for the country as a whole, the main effect of Alaskan oil will probably be to retard the rate of decline of production. The world's peak production may be expected at some time around the turn of the century. The industry must come to terms with these facts, because in every aspect it has been geared to an annual increase in production of at least 5 percent. The situation for an industry which must count on phasing itself out at that rate is entirely different.

In the face of many unknowns, including the absolute amount of oil, future political and economic conditions, and unforeseen breakthroughs in technology, M. King Hubbert describes the situation for oil in the following way.[1] The amount of an exhaustible resource that is ultimately produced cannot exceed the amount initially present, and every drop of oil that is withdrawn decreases the amount that remains. The production of an exhaustible resource is described by the curve in figure 5-2, a. This is a normal growth curve, in which each increment of growth is determined partly by the current rate of growth and partly by the ultimate size that will be reached.

The cumulative amount of production always lags behind the cumulative amount of discovery, which is a fancy way of saying you have to find the oil before you can pump it out. The amount of proved reserves follows a course determined by the relationship between production and

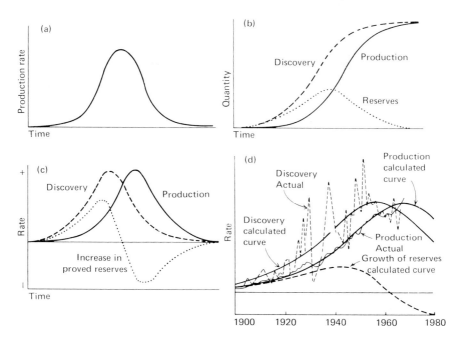

Figure 5-2. Relationships among discovery, production, and reserves of a depletable resource.
a, Theoretical production rate as a function of time.
b, Cumulative quantities discovered, produced, and identified as reserves.
c, The relationship among rates of production, discovery, and addition to reserves as a function of time.
d, Actual discovery and production rates plotted against the theoretical curves.
(Redrawn by permission from M. K. Hubbert, "Energy Resources," in *Resources and Man*, National Academy of Sciences-National Research Council [San Francisco: W. H. Freeman and Co., 1969], pp. 168, 172, 173, 178.)

discovery. Cumulative discovery and production tend toward the same maximum value (assuming that all the discovered oil is produced), and reserves fall to zero (figure 5-2, *b*). The rates of discovery and production first increase, then decrease, until they reach zero, as shown in figure 5-2, *c*. This figure also shows the path followed by the change of proved reserves.

The curves in figures 5-2, *a*, *b*, *c* are theoretical, but if we plot actual data on graphs like these, the correspondence between theory and experience is striking (figure 5-2, *d*). One more example will illustrate the ephemeral nature of our oil binge. The typical growth rate of crude oil production in the continental United States exclusive of Alaska has been 5.86 percent per year. Using the best current estimates of the total amount of oil available and assuming that the rate of production will follow the normal type of curve shown in figure 5-2, *a*, we discover that 80 percent of America's oil will be produced during the sixty-five year period between 1934 and 1999.[2]

The future will see increased emphasis on more complete recovery of oil from oil fields. In the past, with the pump alone, as much as three-quarters of the oil has been left in the ground. With fluid injection, solvent injection, and fracturing techniques, the amount of recoverable oil has reached 40 percent. Some improvement in this figure is probable. The future will also see increased production of oil offshore, from the continental shelves. However, such developments will not change the overall picture. Hubbert's critics say he does not allow for the unpredictability of the future. But the production of oil and other fuels appears to be following a typical growth curve despite the considerable social and technological changes that have already occurred. We could alter the course of production intentionally, but we don't seem to be planning to do it. As for errors in estimates of recoverable oil, we must face the fact that within a wide range, the numerical value of a limit is insignificant if growth persists. If we could double our efficiency of extracting oil (unlikely) and if estimates of oil resources prove too low by 1000 percent (very unlikely), the peak of production would be delayed by little more than half a century. Despite impressive advances in the theory and technology of petroleum exploration, success, as measured in terms of discoveries per foot of exploratory drilling, has fallen in the last thirty-five years from more than 250 barrels per foot to 35. Crude oil supplies are running out.

The story of natural gas is similar to that of oil. Natural gas was a nuisance to the early oil men. They usually allowed it to escape into the atmosphere or burned it on the spot. Only a tiny fraction was salvaged for use as fuel in the immediate vicinity of the oil field.

The problem with putting natural gas to work was twofold. First, there was no known use for it that another fuel could not serve as well. Second, there was no way of delivering it from the gas field to the consumer. The oil industry would be half a century old before its offspring gas could make its own way.

In the late 1920s, development of high-pressure pipe, improved welding techniques, and power equipment for laying pipe made possible the transport of gas at reasonable cost. Meanwhile, because gas and oil often occur together, the gas industry found itself in the strange position of seeing the amount of its product increase automatically as oil production increased, while its market possibilities dwindled correspondingly. As a result, the price of gas remained very low despite relatively high transportation costs, and it is still artificially low today.

Gradually, the advantages of gas as a fuel became apparent. It was a clean fuel, just as the commercials said. It required no storage bins or tanks, but could be piped in as needed, like water. It was burned in simple, inexpensive, almost maintenance-free furnaces. With the growing demand for gas as fuel, its only other major use, for the production of carbon black, declined sharply. Today, natural gas is the most sought-after member of the petroleum family for home and industrial heating and electric power generation. The gas that cooks a roast in New York may have come virtually nonstop through pipelines from Texas. Only parts of New England, lacking the lure of a large market, are beyond the range of gas.

The natural gas of commerce is methane. As byproducts of the production of natural gas, other hydrocarbons chemically related to methane are extracted. Collectively, these are known as natural gas liquids. A fraction of this mixture, composed primarily of propane and butane, is marketed as bottled gas or liquefied petroleum gas for cooking and heating. The remaining heavier fraction is often added to the gasoline fraction of crude oil, which it resembles in composition.

Hubbert has also applied his methods of analysis to the production of natural gas and natural gas liquids.[3] Again, rather than looking at the statistics of production, we will consider the more fundamental question of how far along its path from birth to death the exploitation of this resource has progressed. Figure 5-3 summarizes the story for gas. Although the

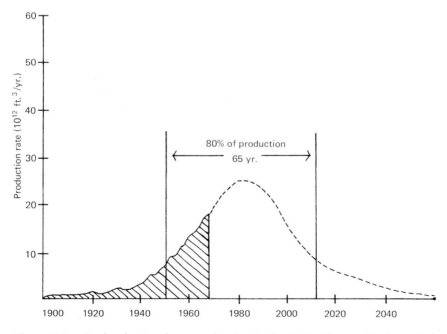

Figure 5-3. Cycle of natural gas production in the United States. (Redrawn by permission from M. K. Hubbert, "Energy Resources," in *Resources and Man,* National Academy of Sciences-National Research Council [San Francisco: W. H. Freeman and Co., 1969], p. 190.)

numbers are different, production of natural gas liquids follows a course similar to oil. As with oil, we see that 80 percent of America's resources will be used up in sixty-five years, the period between 1950 and 2015. Less than a human lifetime. The peak of production may be expected somewhere around 1980; the peak of discovery has already passed. New discoveries are not keeping pace with the growing demand. In the past decade, what was a twenty-year reserve has plummeted to an uncomfortable twelve-year reserve, excluding the resources of Alaska.

Worldwide, the use of natural gas has been hampered by the same transport problems that plagued the United States half a century ago. The wasteful flaring of gas at its source has continued until today, although three recent trends promise to end the waste: an increase in electrical power-generating facilities and gas-consuming industries near sites of gas production, the building of pipelines, and the construction of special tankers for transport of gas in liquid form at low temperatures.

An essentially untapped source of petroleum is tar sands. Tar sands are reservoirs of a heavy type of petroleum, still a liquid, but too viscous to be extracted by conventional pumping. The geological history and chemical nature of this material are similar to those of crude oil. It can be processed with minor modifications in current refining equipment and techniques, but its physical properties create stubborn problems in its recovery from the earth. For this reason, known deposits have only recently been exploited, and so far on a tentative basis. For this reason too, there has been little effort to conduct a world survey of tar sand resources. They represent an enormous potential supply of energy, however. The tar sands of western Canada, which are the best known deposits of their kind, may hold almost twice as much oil as was originally contained in the conventional crude oil deposits of the continental United States.

With present technology, the oil-bearing rock must be mined much as coal is mined. The people who live in the province of Alberta may object to having more than 16,000 square miles of their land dug up, chewed, and spit back. They may object even more to having it done to run the motor boats and air conditioners of their neighbors to the south. The mining has begun. Since 1967, oil from the Athabasca tar sands has been trickling into the United States. So far the amounts are negligible. Total production is about 50,000 barrels per day, only 0.3 percent of the 14.7 million barrels consumed daily in the United States. But the industry plans to push ahead until a maximum production rate is reached in the 1990s.

Development of methods for extracting hydrocarbons from tar sands while leaving the rock in place is almost certain. One promising method is to inject steam into the rock and drive out the hydrocarbons in a steam emulsion. However, about 17 percent of the fuel obtained in this way is needed to produce the steam used in obtaining the fuel. Another method is underground burning, with air injected at one well to support combustion and drive hydrocarbon vapors toward another well, the producing well. This technology will probably improve, if there is incentive. But North America needs oil now, and to that end, a quarter of a million tons of Alberta's soil and rock are dug up each day to produce a paltry 50,000 barrels of oil.

By far the largest supply of fossil fuels is in the form of coal and the related oil shale. Unlike petroleum, coal revealed the secret of its origin long ago, through the many fossils that are beautifully preserved in it. Between 440 and 400 million years ago, plants began tentatively to invade the land. They were simple at first. Very few of their direct descendants survive today. But they were successful pioneers, modifying land and atmosphere and preparing the way for the lush growth to come.

There was a period of mountain building when the seas shrank and continents became dry. The first amphibians ventured ashore. Then the seas came back, spreading over much of the continents and creating vast areas of warm, swampy lowlands. It was about 345 million years ago, the beginning of the great fern forests, the time of the first seed plants, and the start of the greatest period of coal formation the world has known.

Psilopsida, Lycopsida, Sphenopsida. These are strange names of

Reconstruction of a prehistoric coal forest. (Photograph of a diorama, courtesy of Field Museum of Natural History, Chicago, Ill.)

strange plants of that time. They have either become extinct or been so greatly diminished in size and number that you may never have noticed them. There is the whisk fern, not a fern at all but one of three remaining species of Psilopsida and the only one in North America. And the little club mosses (not real mosses) that represent what once was a large and diverse group, the Lycopsida. Of the Sphenopsida, only horsetails, *Equisetum*, remain to remind us of giant plants that once grew 50 feet tall in bamboo-like groves. Evolution has been kinder to the ferns, although the ferns of today are only a faded memory of the past. Enormous cycads, first plants to produce seeds, once flourished — then vanished, leaving their chapter in the earth's diary of rock.

In the warm, humid climate of the coal swamps and forests, plant growth far exceeded losses to predation and decay. Plant material falling into stagnant swamps underwent partial decay probably resembling the process of humification that goes on today. During humification, gases, chiefly carbon dioxide and methane, are driven off and water is removed from organic molecules, leaving a residue that is enriched in carbon. This dark brown, jelly-like humus soaks into less destructible fragments of woody tissue, often preserving details of cellular structure with great clarity. Humified material and other plant remains collect to form peat. The deposits continue to be gradually enriched in carbon, as more methane, (in this context known as marsh gas) is driven off. Burning methane, its pale ghostly flames flickering over a bog, is known as the "will-o'-the-wisp."

Peat forms in swampy lowlands and water-logged uplands, in cool humid regions where temperature inhibits decay and tropical swamps where plant growth outdistances decay. Shallow lakes in glaciated areas, flood plains, and deltas are often converted to peat bogs in the normal course of ecological succession. Study of borings made through the delta of the Ganges River reveals a sequence of layers of peat, sand, and clay, indicating repeated uplifting and subsidence of the land. These conditions are thought to resemble the climates and geographic conditions under which the great coal seams of the past were formed.

The pressure of accumulating sediments and the accompanying increase in temperature drive off more and more water and gases from the peat. When almost nothing but carbon remains, about 1 foot of peat has been compacted into less than 1 inch of coal. Fossilized spores, pollen, and other plant structures remain to reveal the essential nature of coal. Unlike petroleum, coal gave away the secret of its origin many years ago.

The type of plant residue (predominantly trees or predominantly algae, for example) determines the type of coal, while the stage of chemical alteration determines its rank. The rank of a coal is correlated with the maximum depth of sediment under which it was buried. A high rank anthracite may once have been buried under 3 ½ miles of sediments.

Coal seam minable by strip mining methods. (Photo courtesy of Eugene Cameron.)

Coal of lowest rank, most like peat, is called lignite. It is often of relatively recent origin, less than 150 million years of age. Next comes bituminous coal from 300 million years ago. Then anthracite, coal of the highest rank, hard and brittle, burning with a hot and smokeless flame.

Special types of coal contain less carbon than normal coals of corresponding rank but are richer in hydrogen. Some of them, when ground and heated to around 900° F, yield a mixture of hydrocarbons resembling petroleum. These are the oil shales, toward which an oil-hungry world is casting greedy glances. Oil shale, a finely textured sedimentary rock (not necessarily a shale), also differs from normal coal in that it contains a very large amount of inorganic matter. The organic portion is derived primarily from pollen, spores, and algae, unlike that of normal coals which originate from leafy and woody plant material. Although oil shale usually accumulates at the bottom of shallow lakes, some of it is of marine origin. The oldest known oil shales are 600 million years old.

English monks of the ninth century gathered "sea coales" from rocky shores to burn for cooking and comfort. Within a few centuries, coal mines were scattered over Europe, for fuel wood had grown scarce. Oil shale, too, was first burned in Europe during the Middle Ages. In the nineteenth century, small oil shale refineries produced candle wax, kerosene, lubricants, and gas, using techniques that later would be applied to the refining of petroleum.

In America, feverish expansion had caused such depletion of fuel wood that by 1850, shortages of wood gave impetus to the coal and oil shale industries. Drake's oil well put an abrupt end to exploitation of oil shale, but the use of coal continued to grow. Coal rapidly became the primary fuel for industry and transportation, while wood continued to supply domestic needs. Then, in the first part of this century, coal took over the domestic scene as well. Since World War I, the use of coal has increased with our growing production of electricity. The iron and steel industry continues to require large amounts of coking coal.

The 1920s saw the beginning of the switch from coal to fuel oil for residential heating and industrial use. The rise of the automobile and the conversion of railroads to diesel engines after World War II switched the transportation market from coal to oil. An expanding network of high pressure gas transmission lines enabled gas to steal what remained of coal's residential and commercial markets and to make inroads on the oil market itself. At present, except for its role in the iron and steel industry, the use of coal is almost completely tied to the electrical utilities. The situation is tense, for air and water quality standards and shortages of low-sulfur coal have caused power companies to seek oil, gas, or nuclear fuel for their new installations.

To the business mind, environmental and mine safety legislation has cramped the coal industry unbearably. Low-sulfur coal is mined primarily in the southern Appalachian bituminous fields and the northern Rocky Mountain sub-bituminous and lignite fields. With the major use of eastern coal having been until now for manufacture of metallurgical coke, most of this coal is already owned by or contracted to steel producers at prices about twice as high as utility coal brings. New mines are needed, but environmental concerns interfere, while coal men warn that construction of a new mine requires between two and five years plus a large initial investment, an assurance of at least a twenty-year supply of recoverable coal, and usually a long-term purchase agreement.

The longer term outlook for coal is brighter than the immediate stalemate implies, if we can decrease the occupational and environmental hazards of mining. Coal's possibilities as a raw material seem limited mainly by man's ingenuity. Long after the last oil and gas wells have been sealed shut, liquefaction and gasification of coal may provide the petroleum products to which we are now accustomed. Coal will be the raw material for an ever more diverse petrochemical industry, and through microbiological and industrial processes it may even be converted to food. In view of these possibilities it seems very strange that the major portion of federal funds for energy research (86 percent in fiscal year 1970) is devoted to nuclear fission and fusion, while coal scrambles with a host of other contenders for its share of the rest.

Pollution from our present uses of coal may never be controllable. Liquid and gaseous fuels are more versatile than coal, which is why they were able to steal its market. For these reasons, conversion of coal to other fuels looms importantly ahead. It is thought that the conversion process will cause only minor pollution, and the resulting fuels will be clean. However, production of synthetic fuels from coal will increase destruction of land surfaces through mining and hasten the depletion of coal resources — facts which are often overlooked in projections of coal supply and use.

During the nineteenth century, "coal gas" or "town gas," a byproduct of the coking industry, was widely used in Europe and America for cooking and heating. The gas was a mixture of carbon monoxide, carbon dioxide, hydrogen, methane, and minor constituents. The production and use of town gas, however, has little relationship to modern processes for making synthetic fuels.

During World War I, Germany made aviation gasoline by hydrogenation of coal. Today, South Africa produces most of her gasoline and some other petroleum products from coal. The South African plant liquefies about 2000 tons of coal per day, consuming an additional 2400 tons to run the process. The main products are 3660 barrels of gasoline and 370 barrels of diesel fuel. Synthetic gasoline is too expensive to compete with the natural product in the United States, but improved technology and the uncertain future of petroleum imports and production could change the picture. At the moment, several schemes for liquefaction of coal are at or near the pilot plant stage, but are stymied for lack of funds.

Rising prices for natural gas and improved gasification techniques could make gas derived from coal competitive in price with natural gas before the end of the 1970s. Several processes for conversion of coal to gas are under consideration or have already reached the pilot plant stage. The resulting gas may be almost pure methane, suitable for transmission through pipelines, or it may be a mixture of compounds having less potential energy but well suited for the generation of electricity at the site of production.

The net chemistry of coal-to-gas conversion is a reaction of carbon (coal) with hydrogen to form methane. The usual commercial source of hydrogen is natural gas. It would be absurd to produce hydrogen from natural gas for reacting with coal to produce a substitute for natural gas. But where will the hydrogen come from? For many laboratory and pilot experiments, it has in fact come from natural gas. But there is a more abundant source at hand: water. When finely ground hot coal reacts with steam, a mixture containing carbon monoxide, carbon dioxide, methane, hydrogen, and unreacted water results. The proportions of the various gases depend on the temperature and pressure under which the reaction is carried out. This mixture is then cleaned and perhaps further converted to methane, depending on whether it is to be transported by pipeline or burned onsite. However it is accomplished, gasification of coal requires energy. In one process, about two-thirds of the coal is gasified while the remaining third becomes char, which is burned to generate steam for further gasification.

With technology as it was in 1971, the cost of synthetic methane would have been about twice the cost of natural gas.[4] An equivalent amount of energy in the form of synthetic oil would have cost about three-quarters as much as synthetic gas, or 30 percent more than natural crude oil. The estimates for synthetic gas are probably more realistic than those for synthetic oil, partly because the technology is further along and partly because estimates for synthetic oil are spurred by a competitive desire to have the number come out favorably. In any case, the numbers given are very rough. Unforeseen improvements in technology can be nullified by unforeseen problems; and relative costs are subject to extreme fluctuation as a result of political action here and abroad — changes in oil import policies or pollution standards, for example.

As for oil shale, there are still knotty problems, only some of them technical. There are a host of pre-existing mineral claims on prime oil shale lands in America, some of them more than fifty years old. Legal questions may occupy the courts for so long that they may constitute the main hope of conservationists who want to delay large-scale production of oil shale. If we tried to exploit everything at once in these rocks, which also contain valuable sodium and aluminum minerals, the major portion of oil from the shale could be consumed on the spot to provide energy for the other mining operations — hardly a solution to the nation's energy problems.

So far, technical problems and abundant cheap petroleum have thwarted growth of an oil shale industry in America, but the planning has gone on even though actual mining has not. When our petroleum resources are near exhaustion or when foreign oil becomes unavailable or too expensive, large-scale production of oil shale will probably begin. Already, large industries operate in the U.S.S.R. and mainland China, and there are plans for development in Brazil. A prototype leasing and development program in the Green River Formation of Colorado, Utah, and Wyoming is imminent. It is only a prototype program, but if government and industry did not have intentions of pursuing it further, it would not be undertaken.

We can gauge coal resources with more certainty than petroleum, which often lies well-hidden or deep within the earth. The proof of an oil deposit is in a producing oil well. Even exploratory drilling can give, at best, only a rough indication of how much petroleum may eventually be recovered. Coal, in contrast, is usually found close to the surface. The stratified deposits are continuous over large areas and frequently are visible as outcrops at the surface, thus facilitating accurate surface mapping with the help of only a few widely spaced drill holes. Taking minable coal to be 50 percent of the coal in seams a minimum of 12 inches thick, extending to a depth of 4000 feet or in some cases 6000 feet, a recent United States Geological Survey study[5] estimates the total original minable coal of the United States to be almost 1.5 trillion metric tons, and that of the world to be 7.6 trillion metric tons. (A metric ton is 2200 pounds.) About half of these amounts has been established by exploration and mapping; the other half is inferred on geological evidence but has not actually been determined. In the first thousand years of coal mining, less than 10 percent of this supply has been removed from the ground. However, if present rates of increase in production continue, it will take only about 340 years to use the next 80 percent of the world's coal — the consequences, again, of exponential growth.

By all estimates, the amount of oil held in oil shale is immense. The practical potential, however, varies according to whether you are talking to an oil man, a geologist, a politician, or a conservationist. High grade, accessible oil shales in the United States, those that are at least 10 feet thick and contain 25 or more gallons of oil per ton of shale, might provide 80 billion barrels of oil. This is four times as much as is in Alaska's crude oil deposits, but it is only a fifteen-year supply at current rates of consumption in the United States. Less accessible or lower grade shales containing between 10 and 20 gallons of oil per ton could provide about 2 trillion barrels of oil, enough to keep us going for a couple of centuries at the current rate. These shales are marginally or submarginally producible under present conditions. Suspected deposits and possible extensions of known deposits might yield an additional 33 or 34 trillion barrels. By far the largest part of the oil, some 140 trillion barrels, is in "organic-rich shales," a euphemism for shales whose exploitation as fuel may forever remain economically, energetically, and environmentally impractical. Perhaps the best use of shale oil will be in the petrochemical industry, where the quantities required are insignificant compared to quantities burned as fuel, the market can bear a much higher cost for raw materials, and there is at least a theoretical opportunity for recycling. Thus shale oil may be our future source of dyes, drugs, synthetic rubber and fibers, fertilizers, paint, detergents — even food.

Our current fossil fuel shortages and those in store for the near future are often said to be caused by economic conditions and by our new-found concern for the environment rather than by actual lack of fuel. This does not make them less real. In fact, economic and environmental problems are related to the fact that our highest grade fuel supplies have been exhausted. As exploration becomes more difficult and less rewarding, as the cost of producing fuel from poorer and more deeply buried deposits increases, and as the demand for clean fuel grows, it is possible to imagine a time when, no matter how much fuel remains in the earth, it will cost more than the market will pay to produce it and more energy will be consumed in its production than can be derived by burning it. At present, 7 percent of our annual energy production (20 percent of the energy consumed by industry) is used in the extraction and refining of petroleum. Processes for gasification of coal and *in situ* extraction of hydrocarbons from tar sands require even larger amounts of energy than are currently involved in the production of petroleum.

Although we are using fossil fuels at a rate more than one hundred thousand times as fast as they can be replaced, substantial amounts remain for half a dozen human generations or so. The problem, then, is not a shortage of fuel but the problems of producing and using it. Development of a national energy policy is receiving concerted attention in the government because of the serious political, economic, and environmental problems caused by our growing rate of energy consumption.

Until World War II, the United States was a net exporter of energy, including petroleum. Since 1947-48 we have depended increasingly on foreign oil, despite efforts of the government and oil industry to restrict imports and keep the country self-sufficient in petroleum production. Now we import almost 30 percent of the petroleum we use, most of it from South and Central America, Mexico, and Canada. Although we are the world's largest exporter of coal, the energy value of our coal exports is only about one-fourth that of our petroleum imports. Predictions for the future indicate that we will rely more heavily on foreign oil, with perhaps as much as 40 percent of it imported by 1980 or 60 percent by 1985.

Japan and western Europe are in a worse predicament than the United States, as they are industrialized areas almost totally dependent on oil imported from developing nations — although recent oil and gas discoveries in the North Sea promise to ease things for Europe.

The developing nations realized that they were being taken, for their most valuable resource, which they needed to build a better future, was being sucked away, at bargain prices, into the overdeveloped countries.

The Organization of Petroleum Exporting Countries (OPEC) was formed during the 1960s to combat the economic power of the international oil companies. Since then the OPEC countries have raised the price of oil sharply. When the industrialized countries were shopping for the cheapest foreign oil concessions they called it free enterprise; when the OPEC countries demanded higher prices, some called it conspiracy.

Meanwhile, in the Congress of the United States and among leaders of finance and the energy industries, the question of foreign oil is still debated. Some claim that national security demands that we be self-sufficient, or at least able on a moment's notice to become self-sufficient, in petroleum production. Dependence on foreign oil that can suddenly be withheld or made more expensive on political grounds could cause a paralyzing energy shortage, as events in the Middle East during the last part of 1973 demonstrated. This reasoning lay behind the establishment of our controversial oil import quotas. Others claim that we are pushing the domestic production of a resource we really want to conserve, and at prices which cost the American consumer billions of dollars annually above what foreign oil would cost. We should get as much foreign oil as we can while we can, they say. And then there are those who ask quietly if it is right to demand an end to oil production in the Santa Barbara Channel and let production of gasoline for Santa Barbara's automobiles pollute the coast of Venezuela instead.

While we scheme to increase the availability of fossil fuels we are becoming increasingly aware of the problems related to their use. Environmental degradation accompanies every stage of discovery, production, and use of fossil fuels. Some of the results of our addiction to energy are obvious: oil-smeared beaches, the wasteland of southern Appalachia, belching smokestacks, and traffic jams. Other problems are more insidious.

Damage to the environment from exploratory activities is generally transient and negligible. Exploration in subpolar regions is an exception. The arctic already shows scars which justify alarm over the impact of future activity. Life clings precariously in the far north. Soil and even climate are easily disturbed. A few hundred species of plants (compared to thousands in the tropics), a few dozen species of higher animals, some insects, soil invertebrates, and microorganisms have adapted to the short growing seasons, low temperature, and lack of moisture. Where the ground is permanently frozen, a condition known as permafrost, the cover of vegetation provides shade and insulation which maintain the frozen state. If vegetation is disturbed, thawing occurs. Upper ground thaw kills plants and permits still more thawing. With permafrost and groundcover destroyed, erosion can be rapid and severe. Recovery could take centuries.

Prospecting in the north requires the transport of equipment and people. We have learned through our mistakes how to build roads that hold

Erosion across tundra near Canning River area of Alaska's North Slope. This deep scar was made by a crawler tractor in the process of setting up a seismic exploration for petroleum. (Photo courtesy of Bureau of Land Management, U.S. Department of the Interior.)

environmental disruption to a minimum, but this requires care and we are not always careful. The mere presence of people places a severe strain on the arctic environment. As one ecologist said, "You can't just bury your garbage."

More serious problems are related to the production of fossil fuels. For more than 5000 years, man has tunneled into the earth to gain buried treasure. About 1000 years ago the mining of coal began. The pick and shovel of the coal miner, his hand drill and mule cart, were the same as those of any other subterranean digger. There were problems with pollution of streams and groundwater and with the ugliness of the mines themselves; with long-burning underground fires and subsidence of the surface; and with sickness, injury, and death among miners. But these seemed a small price to pay for the energy that would perfect the world.

The true cost of underground mining is revealing itself today. When rock that has lain buried and undisturbed through geologic ages is broken up and exposed to air and water, chemical changes occur. Sulfide from the ubiquitous iron pyrites (fool's gold) is oxidized to sulfate and sulfuric acid. Entering the groundwater, the sulfuric acid causes "acid mine drainage." Many materials that are relatively insoluble in less acid water dissolve under the conditions of the mines. These include iron, magnesium, manganese, copper, zinc, and calcium. When the drainage from mines pollutes water supplies most aquatic plants are killed, the water can no longer support fish life, the water becomes unacceptable for recreational uses, and costs of treatment for municipal water supplies are increased.

In some densely populated areas of Pennsylvania, the land is sinking into a maze of forgotten tunnels, some of them mined out and abandoned a century ago. Expensive as subsidence control projects are, post-subsidence repair can cost up to ten times as much. There are some sixty thousand abandoned coal mines in the United States. They can be filled, and foundations of the buildings standing upon them strengthened, but acid water problems remain. The mines can be sealed off at a cost of up to twenty thousand dollars per mine, and the water can be treated — at considerable cost and with variable success. The basic problem remains. The scale of mining forces the earth to come to a new balance, and the balance is not to our liking.

About fifty years ago, strip mining began. Powerful earth-moving machinery easily and economically scraped away surface layers of soil and rock, permitting recovery of 90 percent or more of the underlying coal. But by the 1950s the penalties for stripping became apparent, and Appalachia became the battle cry for social and environmental reformers.

Material stripped from the surface and piled to one side or pushed down a mountain invites erosion by wind and water. Sediment washed from strip-mined slopes often clogs stream channels, destroying fish habitats and subjecting adjacent land to flooding. It covers highways and settles on farmland, killing crops. Costs of water treatment for municipal and industrial supplies increase. Reclamation, especially in mountainous or agricultural areas, becomes a myth perpetrated by the public relations experts at the coal companies. In arid regions, reclamation is frankly admitted to be impossible. As in underground mining, water supplies become contaminated with acid and high concentrations of dissolved salts.

Strip mining for coal has occurred in about half of the states in the United States. It provides almost half of the coal on the United States market today. It has not yet touched Alaska, where vast coal seams lie shallowly buried. The largest earth-moving machine can chew up more than 4,000,000 cubic yards of soil and rock in a month. It is feasible to remove surface material to a depth of 200 feet in order to get at a coal

Above, an airphoto of an active strip mine in Montana. This mine is subject to the Strip Mining Reclamation Act of 1973. The area on the left is being regraded, but the scars will remain for some time. *Below,* a 24 cubic yard shovel used in strip mining. The Volkswagen parked in front of the machine gives an indication of its tremendous size. (McKelvey photos, courtesy of Montana Department of State Lands.)

seam. Some 2500 square miles of land in the United States have already fallen to the bulldozer and power shovel, and projections indicate that by 1980 this number could reach nearly 4300 square miles — the equivalent of four-fifths of the area of Connecticut. It is a mere 6 percent of the total area thought to overlie coal recoverable by stripping.

The cumulative environmental debt of coal probably exceeds 4 billion dollars in the United States.[6] This debt will grow before it shrinks. Necessary as figures like this are, it seems absurd to give people money for the misery they have suffered and say the debt is canceled.

Insofar as they must be mined, the exploitation of tar sands and oil shale presents the same problems as the mining of coal. In addition, for oil shale, there is the problem of disposing of spent shale after processing; for the volume of waste material is greater than the volume of material originally mined. The president of Atlantic Richfield suggested filling in some of the surplus canyons and valleys in uninhabited areas of the west, but there is little applause for that idea. The first extensive production of oil shale will occur in 17,000 square miles of the Rocky Mountains' Green River Formation, a dry semiwilderness region where, as in the arctic, the environmental impact of people will probably be worse than the impact of actual production. A prototype program calls for employment of eighteen thousand workers, accompanied by their families, automobiles, sewage systems, power plants, and so on.

Rich oil shales in northwestern Alaska promise up to five times as much oil as the Prudhoe Bay oil field just to the east. Production of this shale would pose problems of gargantuan scale and would bring environmental ruin.

"See the Oil Wells in the Sea," beckoned an advertisement for the Southern Pacific Railroad's Los Angeles-Santa Barbara run when the first offshore wells were sunk off Summerland in the 1890s. Several famous gushers of the past have attracted attention from tourists, while more recently the Santa Barbara oil spill lured sightseers. Most people, however, do not like to see oil wells.

When all goes well, drilling for oil is not particularly disruptive. The leakage from normal operations is negligible compared to the total efflux of oil into the environment. Routine offshore production accounts for only 4 or 5 percent of the oil that pollutes the sea. But there are accidents. Offshore blowouts are always potentially disastrous because coastal ecosystems are especially sensitive to oil and because it is difficult to control, contain, and clean up a spill at sea. Yet all the offshore accidents, spectacular and unsung combined, contribute less than 5 percent of the petroleum that we dump into the sea.

Onshore spills are easier to deal with. The arctic, again, presents special problems, for the effects of oil pollution are most severe at high latitudes and in cold weather. Spilled oil is eventually degraded by chemical weathering and the action of microorganisms. The rates of these processes are influenced by the temperature, the general rule being that the rate of a chemical reaction is roughly cut in half for each 14 to 18 degree drop in temperature. In addition, microbial activity in the arctic is limited by lack of nutrients, so oil spilled in the arctic will stick around for a long time.

Perhaps the worst consequence of oil production is that it spawns storage, refining, and transport facilities, which are the real villains of oil pollution. It is not the sight of a few offshore platforms that alarms residents along the Atlantic seaboard. It is the anticipation of surrendering the coast to ceaseless tanker traffic and belching refineries.

After fossil fuels have been removed from the ground they may still be a long way from their end use. In the past, little was done to treat coal except for crushing and sorting by size and perhaps cleaning. Today about half of the coal produced in the United States is cleaned, a process which pollutes water, adds three-quarters of a million tons of particulate matter to the atmosphere, and creates 90 million tons of solid waste each year.

Petroleum, on the other hand, has always been treated extensively in order to derive a variety of end products. Our comments on refining of petroleum apply equally to oil shale and tar sands.

There is nothing subtle about the pollution from a petroleum refinery. You can see and smell it. Each year the petroleum refineries of the world spew more than 12 million tons of carbon monoxide, about 3.5 million tons of sulfur oxides, and almost 11 million tons of gaseous hydrocarbons into the air, as well as about a million tons of particulate matter. On a weight basis, they are the largest industrial polluter, although almost half of the material is carbon monoxide, least troublesome of the major air pollutants. Water pollution is also a problem. Refinery effluents contain ammonia, phenols, naphthalene, cresols, sulfides, and other noxious chemicals.

Many refineries are located along the seacoast, where crude oil arrives by tanker and refined products depart. Through normal operation and accidents, these facilities contribute about 15 percent of the oil with which we pollute the sea. An estimated fifteen thousand oil spills each year are attributed to land-based facilities, many of which are refineries. About 83 percent of these accidents pollute the sea and another 8 percent foul inland lakes and rivers.

There is an additional possibility of environmental damage in the transport and storage of fuels. Transportation and storage of coal create no special problems except aesthetic ones. Increasing numbers of "unit" trains consisting entirely of coal cars shuttle back and forth between mine and power plant or steel mill. A few pioneering coal pipelines are in operation. Coal deteriorates on storage and there is a possibility of spontaneous fires and water pollution; however, proper management prevents these problems.

Above, the Santa Barbara oil spill in 1969, about three weeks after the initial blowout. The two largest leaks are marked by the bubbling gas escaping along with the oil. *Below,* a closer view of platform A in the Santa Barbara channel shows the leaking area at the corner of the platform. The dark area at the top of the picture is covered by little or no oil, in contrast to the thick accumulations (lighter area) occupying most of the picture. (Photos courtesy of Marine Studies Center, University of Wisconsin.)

Petroleum and natural gas are moved chiefly by pipeline and tanker. About 70 percent of our domestic crude oil and gas flow under high pressure through 200,000 miles of pipelines. Accidents are frequent. The federal government has only recently begun to collect data in this area and the earliest information is incomplete. Data from 1970 suggest that the accident rate may be well over one thousand per year. Pipeline leaks account for about a quarter of all offshore spills.

Most leaks are small, but there are a few dramatic ones, like the break in Lima, Ohio, in 1969, which poured 700,000 gallons of crude oil into the Ottawa River. The danger of an oil pipeline spill is mainly to wildlife, while a gasline leak raises the spectre of fire, explosion, injury, and death in populous areas. It is possible to reduce but never to eliminate damage from these accidents through proper surveillance and safety devices, including automatic monitoring and closely spaced shutoff valves. But with perhaps 250,000 gallons of oil in each mile of pipe, even with safety locks every 10 miles, 4 million gallons of oil could escape during a pipeline leak.

Since the discovery of oil in Prudhoe Bay, a word association test would probably produce one response with regularity: "pipeline-Alaskan." The difficulties and dangers of such a pipeline are legion. Underground pipe carrying hot oil could melt the permafrost, with consequences ranging from minor to severe. Insulating the pipe would raise the temperature of the oil, not reduce melting; and if the oil were not hot, it would not flow. Muddy slush in the thawed area could flow down the slightest slope, damaging the pipeline as well as the local environment.

Permafrost is not the only problem. Animal ecologists fear that herds of migratory game will be disturbed if their paths cross that of man. There are extensive fault zones and areas of seismic activity. The act of laying a pipeline could cause more disturbance to the environment than the presence of the pipeline itself — a somewhat specious argument used by the oil industry, since you can't have a pipeline if you don't first construct it. Original plans for the Alaskan pipeline called for the state to take over the service roads after the pipeline was completed — a public road into the backland!

Overland, oil moves through pipelines, but much oil must cross the water. In 1967, the wreck of the tanker *Torrey Canyon* off the coast of England shook the world. The next year it was *Ocean Eagle*, near San Juan, Puerto Rico. Accidents like this are now so common that they are no longer big news. More than half of the world's annual production of petroleum, about 300 billion gallons, is shipped in a growing fleet of about four thousand tankers. It crisscrosses the seas, touching every major island and coastline. Through spillage or leakage, an estimated 0.1 percent of this transport — more than a quarter of a billion gallons — is lost each year to the marine environment. This equals the annual marine production of petroleum-like materials from natural processes like submarine oil seeps and decay of organic matter.

This supertanker, at 252,000 tons, is about ten times as large as the largest tanker afloat in 1950. Supertankers now under construction will dwarf even this ship. (Photo by Ralston — Texaco, Inc.)

With the number and size of tankers growing year by year, we can expect nothing but more accidents and more pollution. Capacity of the world's tanker fleet doubled during the 1960s and will probably double again during the 1970s. The once-supertanker *Torrey Canyon*, at 120,000 tons, is dwarfed by the 800,000-ton ships now on the drawing boards. These super supertankers pose problems in addition to pollution. Only four ports in the United States can handle the 60-foot draft of a 200,000 ton tanker. Colossal dredging projects will be necessary to deepen and maintain shipping channels to accommodate the mammoth new tankers that are thrusting themselves into the seaports of the world.

Oil tankers are notorious polluters in their routine operation. "Normal" tanker activity accounts for one-quarter of man's discharge of oil to the oceans, five times as much as is contributed by major spills or offshore production. It involves both the disposal of oil from boiler and engine rooms (a problem common to all shipping) and disposal of wastes from the cargo. The problem of cargo wastes is being decreased by improvements in the design and operation of tankers and by efforts to reprocess waste oil. However, the immense world fleet will not be retired overnight to be replaced by cleaner ships, nor will the political, economic, and institutional difficulties of salvage be worked out easily.

In many tankers, seawater ballast is taken into the same tanks that carry oil. When the ballast is drained, there is always a residue of oil, or worse yet, an oil-water emulsion. Who is responsible for disposing of this waste? The owner of the tanker? The owner of the cargo that created it? The owner of the receiving terminal? The owner of the next cargo, for whose load the waste must be discharged? Waste oil is usually claimed by nobody and often dumped into the sea.

Recent development of refrigerated tankers for transport of natural gas has greatly expanded the market for this fuel and has not yet posed any serious problems.

Storage facilities for petroleum and natural gas are subject to damage from earthquakes, sometimes leading to fires like those that raged in Seward, Alaska, after the earthquake of 1964. Transfer of petroleum to and from storage tanks is an additional source of spillage and pollution.

The effects of spilled oil would be less noticeable and less severe if the oil were distributed uniformly over the seas, but it is not. It is concentrated in coastal regions which also receive a disproportionate share of man's other wastes and which constitute a particularly intricate, valuable, and vulnerable system of living organisms and physical processes. The largest part of marine biological resources occurs in coastal areas. Moreover, tanker traffic is concentrated in particular lanes, so that, for example, about one-quarter of the world's production of oil passes through the English Channel.

Oil can kill in three ways. It can coat an organism or its respiratory surfaces, causing death by suffocation. It can kill through one of its deadly constituents, largely water soluble compounds of low molecular weight like phenol, that are acutely toxic to many plants and animals. The third route to death is subtle and insidious, difficult to detect and often impossible to prove. It may lead, paradoxically, to local death of a species while sparing the individual. It is the effect of chronic pollution: impaired physiology, genetic aberrations, cancer, decreased fertility or sterility, reduced growth, increased susceptibility to disease and other stresses, or disruption of chemical communication leading to abnormal feeding, defensive, reproductive, and migratory behavior.

Chronic oil pollution may also lead to depletion of oxygen in the water through an excessive increase in the biochemical oxygen demand. The biochemical oxygen demand is a measure of the oxygen needed to oxidize all the organic matter present. Oxygen is depleted by the respiration of a large population of bacteria feeding on petroleum and of the protozoa that feed on the bacteria. Thus both direct and indirect effects of oil cause changes in the number and kind of species in polluted water.

Although many aquatic plants are relatively immune to oil, the widespread use of petroleum products for weed control in irrigation ditches and elsewhere testifies to their potential powers. The first effect of oil is to kill many plants that are rooted in bottom sediments, but when revegetation takes place, growth is likely to be more luxuriant than before because animal predators have been destroyed and because the decomposing petroleum is excellent fertilizer. Accumulation of petroleum sludge alters the properties of bottom sediments and plays havoc with growth and reproduction of bottom-dwelling organisms.

There are few records of the effect of oil on aquatic mammals. The experience in Santa Barbara suggests that it is minimal for seals, sea lions, whales, and dolphins. Fuel oil is known to destroy the waterproof qualities of muskrat fur.

Waterfowl suffer acutely from oil pollution. A single oil spill can kill thirty thousand birds. Many experts believe that a far greater toll than this is exacted slowly but inexorably far out at sea, where the passing of the victims goes unnoted. Oil sticks to a bird and he can't remove it. It destroys his buoyancy so he must struggle to stay afloat. It also destroys his natural insulation, exposing him to cold air and water. It mats his feathers so he can't fly. In preening, he swallows toxic oil. As a result of all this, he has trouble finding food. Suffering from exhaustion, exposure, sickness, and starvation, he has a slim chance for survival.

Pollution associated with the end use of fossil fuels is familiar to everyone. Although many kinds of processes lead to pollution of one sort or another, there is only one major cause of air pollution: burning. About 80 percent of all air pollutants are generated by the burning (and steps that lead to burning) of fossil fuels, and most of the rest comes from the burning of something else. Table 5-1 shows the sources of major air pollutants.

Coal is dirty. Both pollution and evolution are documented by the darkening of certain moths in industrial regions of Great Britain. Initially, these moths were pale animals that avoided predators by sitting on pale objects. The occasional dark mutant was quickly spied by a hungry bird. In the cities, however, there was nothing pale to sit on. Dark mutants multiplied unmolested, while wild types had little chance for survival.

In one hour's operation a typical coal-fired electric power plant producing 1 million kilowatts of power also produces, from 340 tons of coal: 940 tons of carbon dioxide, 13 tons of carbon monoxide, 17 tons of sulfur oxides, 3.4 tons of nitrogen oxides, and 34 tons of ash.

Most of the particulate matter (ash) that tries to escape up the stack can be trapped by various devices. The most effective of these are very expensive, and their operation consumes as much as 5 percent of the electricity produced. The 1 percent or more of particulates that evade the best efforts at capture blow about creating dirt, causing or exacerbating respiratory problems, reducing the amount of sunlight that reaches the

Table 5-1.   Sources of Air Pollution in the United States (1968)

Source	Carbon Monoxide 100 x 10⁶ tons 63.8%	Sulfur Oxides 33 x 10⁶ tons 2.4%	Hydrocarbons 32 x 10⁶ tons 51.9%	Nitrogen Oxides 21 x 10⁶ tons 39.3%	Particulates 28 x 10⁶ tons 4.3%
Fuel burning for transportation	1.9	73.5	2.2	48.5	31.4
Fuel burning in power plants, home furnaces, and other stationary sources					
Industrial processes other than fuel burning, including petroleum refining	9.6	22.0	14.4	1.0	26.5
Solid waste disposal	7.8	0.3	5.0	2.9	3.9
Forest fires, coal waste fires, agricultural burning, gasoline marketing, etc.	16.9	1.8	26.5	8.3	33.9

(Reprinted, by permission, from J. Holdren and P. Herrera, *Energy* (San Francisco: Sierra Club. 1971), p. 145.)

earth, and acting as nuclei for local fog. Increasing the height of the stack and decreasing the size of the particles merely disperses the particles higher, wider, and more permanently.

Only the inorganic form of sulfur (mostly pyrites) can be removed before burning. The rest, about half of it, is bound in organic compounds. Although theoretically sulfur oxides can be removed from stack gases, no existing method is satisfactory. For this reason, many standards for air quality specify a maximum permissible sulfur content of fuel before burning, rather than a maximum level of sulfur emission.

Nitrogen oxides persist in the atmosphere for only a few weeks. They are rapidly dispersed by winds, washed down by rain, or incorporated into living matter. Before this happens, however, they participate in a series of photochemical reactions that lead to smog. They also cause cumulative respiratory damage which takes the form of emphysema or greater susceptibility to other respiratory problems. It has been suggested that eutrophication of lakes near urban areas is hastened by nitrogen oxides precipitated with rain.

Advanced technology promises to reduce most of these pollutants, but this is still in the future. Experiments are slow because of their large scale and great cost. Even with the new technology, it will be many years before existing coal plants of traditional design can be phased out.

Many power plants burn residual fuel oil, the dregs of the refining process. These plants produce about the same levels of sulfur and nitrogen oxides and carbon monoxide as coal-fired plants.

The major portion of carbon monoxide is produced by the inefficient internal combustion engine, making it more of a hazard than the same amounts would be if spewed high into the air from a smokestack. The half-life of carbon monoxide in the atmosphere ranges between one month and five years, which means that high local concentrations can build up unless dispersion mechanisms are efficient. Theoretically, air movement should be more than adequate to ventilate the most crowded city streets, but cleansing winds cannot reach small pockets of pollution and the deep chasms among city buildings. The efficiency of gasoline combustion decreases with start-stop driving so that an idling engine produces about 2 ½ times the pollutants of one operating at cruising speed. Thus, both driving conditions and the number of cars in cities lead to enormous rates of carbon monoxide emission. Improvement brought about by emission controls is offset by increased numbers of vehicles and less than optimal functioning of pollution control devices.

The city dweller breathes air containing, on a year-long average, about 8 parts per million of carbon monoxide, with the level sometimes reaching 50 parts per million. Only 1 part in a million produces detectable physiological effects. Carbon monoxide competes vigorously with oxygen for binding sites on red blood cells and clings tenaciously once it has found a site. Thus it accumulates in the blood during prolonged exposure. It is not like the pollutants that burn your eyes or make you cough. It makes you just a little less alert, so that perhaps you don't stop quite soon enough to avoid that other car. And of course every molecule of carbon monoxide represents wasted energy, the energy to be derived from burning carbon monoxide to carbon dioxide.

Hydrocarbons enter the atmosphere largely through evaporation, but also through incomplete combustion. At least 2.5 percent of our annual production of gasoline, more than a billion kilograms, is lost by evaporation during transfer from refinery to truck to holding tank to car and during standing in reservoirs and carburetors.[7] The loss may be as high as 15 percent of the gasoline marketed.[8]

Hydrocarbons in the atmosphere undergo a series of chemical reactions which result in smog. Some hydrocarbons are potent carcinogens. Rain brings many of the molecules back to the ground, where they are metabolized by soil bacteria and do no further harm. Others enter rivers or lakes or the sea, where they join industrial hydrocarbon wastes and used crankcase oil in adding to the burden of hydrocarbons in the hydrosphere. All of them represent energy which has eluded us, only to return to cause us sorrow.

A home furnace in good condition is a relatively efficient device, but it is not pollution-free. Space heating with oil and gas produces annually 1 million tons of nitrogen oxides, 1 million tons of hydrocarbons, 1 million tons of particulate matter, 2 million tons of carbon monoxide, and 3 million tons of sulfur oxides.

References

1. M. K. Hubbert, "Energy Resources," in *Resources and Man*, National Academy of Sciences-National Research Council (San Francisco: W. H. Freeman and Company, 1969), pp. 170-184.
2. *Ibid.*
3. *Ibid.*
4. H. Hottel and J. Howard, *New Energy Technology* (Cambridge, Mass.: MIT Press, 1971).
5. P. Averitt, "Coal Resources of the United States," *U.S. Geol. Survey Bull.* (1967): 1275.
6. G. Garvey, *Energy, Ecology, Ecomony* (New York: W. W. Norton & Co., Inc., 1972), p. 88.
7. E. Goldberg, "The Chemical Invasion of the Oceans by Man," in *Global Effects of Environmental Pollution*, ed. S. F. Singer (New York: Springer-Verlag, 1970).
8. G. Garvey, *Energy, Ecology, Economy* (New York: W. W. Norton & Co. Inc., 1972), p. 122.

6

NUCLEAR ENERGY

*The problem of transmutation and the liberation of atomic energy
to carry on the labour of the world is no longer surrounded with
mystery and ignorance, but is daily being reduced to a form
capable of exact quantitative reasoning. It may be that it will remain forever unsolved. But we are advancing along the only road
likely to bring success at a rate which makes it probable that one
day will see its achievement. Should that day ever arrive, let no one
be blind to the magnitude of the issues at stake, or suppose that
such an acquisition of the physical resources of humanity can be
safely entrusted to those who in the past have converted the
blessings already conferred by science into a curse.*

Frederick Soddy, 1920

"And just at that instant there rose as if from the bowels of the earth a light
not of this world, the light of many suns in one, . . . a great green supersun
climbing in a fraction of a second to a height of more than eight thousand
feet, rising ever higher until it touched the clouds. . . . Up it went, changing colors . . . from deep purple to orange, expanding, growing bigger, . . .
an elemental force freed from its bonds after being chained for billions of
years. . . . Then out of the great silence came a mighty thunder. . . — the
first cry of a newborn world."[1]

Thus does science writer William L. Laurence describe the explosion
of the first atomic bomb in the New Mexican desert on July 16, 1945 —
"first fire ever made on earth that did not have its origin in the sun." Many
discoveries, combinations of serendipity and calculated experiment, mark
the amazing trail to man's use of nuclear energy. The Atomic Age has
many birthdays and many parents.

The immense power represented by this nuclear test in the Pacific has filled some with optimism about a new, limitless source of energy. Others see nuclear power as a threat. (Photo courtesy of Atomic Energy Commission.)

In the nineteenth century, as Queen Victoria aged, much of the Western world seemed to age with her. The spirit of revolution and romanticism with which the century had begun gave way to the manners and moralities of Thackeray and Trollope novels. The spirit of those times has come, in our day, to be considered the epitome of the ordered and repressed life. It was the Victorian age.

In the rarefied world of physics, giants like LaPlace and Maxwell were finishing the work begun two centuries before by Isaac Newton. By the 1880s it was commonly heard that physics was nearly finished. The best men of the times were busy refining measurements and explaining details. The universe was at last understood as a vast machine whose minutiae needed only to be worked out, to describe both the past and the future with a mechanical simplicity. Even the violent notions of Karl Marx caught the spirit of the times.

The crack in this wall came from an unexpected quarter. In 1895, an undistinguished middle-aged physics professor named Wilhelm Roentgen was experimenting with a cathode ray (electron) tube, when he noticed that something emitted from the tube caused fluorescence of a platinum-containing screen that lay nearby. The mysterious something was called X rays. Its discovery precipitated a flurry of experimental activity, until within weeks, surgeons were viewing X ray images of their patients' bones while Victorian ladies blushed.

Meanwhile, in Paris, Henri Becquerel was investigating the relationship between X rays and fluorescence. He was searching for a fluorescent material that would give off X rays when exposed to ordinary sunlight. With a large collection of fluorescent substances at his disposal, he chose a salt of uranium for his studies. After exposing the uranium sample to sunlight he placed it against a photographic plate wrapped in heavy black paper. When he developed the plate, he found that it had been blackened — surely by X rays, since whatever it was had penetrated the thick paper. An overcast sky prevented the impatient Becquerel from repeating his experiment, so he put an unexposed photographic plate, carefully wrapped, in a drawer and laid the uranium on top of it. Days later, he developed the plate to see whether there had been any residual activity in the uranium. The plate was even blacker than the first had been! Further experiments revealed that the radiation from uranium was totally unaffected by light or by any other conditions that Becquerel could devise. Becquerel had discovered radioactivity.

It soon became evident that the uranium-bearing ore, pitchblende, contained more radioactivity than could be accounted for by uranium itself. In solving this mystery, Marie and Pierre Curie discovered two more radioactive elements, which they named polonium and radium.

Meanwhile, basic notions of the atom itself were being revised. In 1897, J. J. Thomson discovered electrons and correctly deduced that they arose from molecules of gas that became ionized, or positively charged.

The negative electron must then be a constituent of the atom — which was not a billiard ball after all, but something with a structure. Physicists around the world began to construct hypothetical models of that structure and to test the models experimentally. Radioactivity was to be their key to the puzzle.

Rutherford showed that one type of emission from radioactive material, called alpha radiation for want of a better name, was identical with atoms of helium stripped of their two electrons. He next used alpha rays to probe atomic structure. By analyzing the deflection of the relatively heavy, positively charged alpha rays when they passed through thin gold foil, he concluded that the positive charge and most of the mass of the atom are concentrated in a space no more than one ten-thousandth of the diameter of the whole atom. The atom was mostly empty space.

Rutherford continued to shoot alpha rays at various things, in search of experimental data on nuclear structure. His most sensational discovery was that the emission of radiation is accompanied by a profound chemical change, the transformation of one element into another. Alchemy had died hard. Atomic theory had struggled for centuries to assert itself. Now the immutable atom was not immutable after all, but could be changed at the whim of man. Visions of splitting the nucleus titillated the brains of physicists, while the young Houtermans boldly suggested that the reverse process, nuclear fusion, powered the stars. The sober reflections of Frederick Soddy, Rutherford's assistant, are expressed in the quotation at the beginning of this chapter.

In 1932, upon Chadwick's discovery of the neutron, physicists had yet another weapon for their seige of the nucleus. With neutron bullets, Irene Curie, daughter of Marie, and her husband Frederic Joliot created new radioactive elements from stable ones. In Rome, Fermi began a systematic study of the elements, bombarding each in turn with neutrons and characterizing the products. By the time he reached uranium, its 92 protons and 146 neutrons constituting the largest nucleus in nature, a pattern had emerged. The products of a nuclear reaction were neighbors of the parent element, often having one or two more protons than the original nucleus. When he reached uranium, Fermi was hoping to create brand new elements — previously unknown transuranium elements.

He calculated that such transuranium elements would emit strong radiation of a particular range. To detect it, he used a thin strip of aluminum foil to filter out shorter range radiation. This filter prevented him from realizing that something very strange was happening. When an array of elements appeared as a result of bombarding uranium with neutrons, Fermi confidently announced that his goal of creating elements beyond uranium had been reached.

In Germany, Ida and Walter Noddack had been searching for natural transuranium elements for years. As accomplished chemists, they showed that Fermi's chemical analyses failed to substantiate his conclusions. Ida Noddack suggested that, instead, Fermi had broken the uranium nucleus into a number of large fragments. Her peers ignored her because it seemed kinder than ridiculing her. Every self-respecting physicist knew that only particles of enormous energies not yet even approached by man could split the atom.

But peculiar things continued to happen. In Zurich one day, investigators forgot to place the aluminum filter in front of their apparatus and they saw wild oscillations on the screen. Concluding that the instrument was at fault, they replaced it, and this time they remembered to put the filter where it belonged. This event was repeated, with variations, in Cambridge, Paris, and Berlin. Through the cloud of their preconceptions, scientists were unable to understand what they saw. Irene Curie correctly identified lanthanum among the products of her experiment, but then retracted her report for technical — and mistaken — reasons. Meticulous chemical detective work by Hahn, Strassman and Meitner identified new radioactive forms of barium, cerium, and lanthanum in the products of nuclear reactions, but their physicist instincts shouted "impossible!" They couched their report in doubts and apologies (all the while fearful that Irene Curie would discover her error and beat them into print), explaining that their conclusions could be deriving from an unlikely combination of mistakes.

But Strassman was a respected chemist and his data were above reproach. When the results became known, investigators in several laboratories hastened to repeat their own experiments. This time, they omitted the filter so that all radiation would be detected. And now, with open minds, they could comprehend what they saw. They saw intense electrical pulses, unlike anything other experiments had ever shown, representing tremendous energies that could come only from the splitting of the atom.

By the late 1930s, the aims of Hitler had become clear. In the minds of some who knew the secret of atomic energy lurked a fear that Hitler might create an atomic bomb. Impossible, said none other than Neils Bohr. Energy cannot be released like that. Atomic energy is only a laboratory tool. He was seconded by Einstein and Rutherford, among others. But then Hahn, Strassman, and Meitner showed that one lazy neutron could split uranium. And on March 3, 1939, Leo Szilard and Walter Zinn showed that in the events following absorption of a neutron by a uranium nucleus, additional neutrons — more than one — are discharged. Those neutrons could enter and split other uranium nuclei, in a so called chain reaction. Although many obstacles remained, scientists could no longer delude themselves. A uranium bomb was possible.

The rest of the story is well documented and well known. Sequestered in secret cities not shown on any map, in an historic race with imagined counterparts in Nazi Germany, American scientists built the first atomic bombs. To their chagrin it turned out that Hitler was not trying to build bombs at all, having been persuaded that the war would be over before the project could be completed. Germany's aborted efforts in applied atomic physics were aimed at obtaining an additional energy resource, a "uranium machine," to supplement inadequate supplies of conventional fuels and power. But America had the bomb, and, perhaps to avoid being charged with a 2 billion dollar boondoggle, the military seemed determined to use it. On August 6, 1945, America dropped an atomic bomb on Hiroshima.

Since then we have had more and bigger bombs, and the Cold War, and the nuclear arms race, and fallout, until the subject of nuclear energy has become so fraught with emotion that it is difficult to confront the facts rationally. Amidst it all there have been rosy fantasies of a utopia which only abundant energy from the peaceful atom can bring. We are getting things into better perspective now. People no longer dive into fallout shelters when the factory whistles blow, but the peaceful atom is turning out to be less peaceful than we had hoped.

The energy of chemical reactions is derived from rearrangements of the outer electrons of atoms. Most of an atom's energy, however, is concentrated in its nucleus — between 1 and 10 million times as much as resides in the outer electrons. At the same time, most naturally-occurring nuclei are very stable, remaining unaffected by the levels of disturbance familiar in the world of chemistry. That is why the nucleus was so long in yielding to man's probes.

Essentially all of the atom's mass is in its nucleus, which is composed of protons and neutrons. The number of protons in the nucleus establishes the chemical identity of an element (uranium, for example, has 92), but the number of neutrons in the nucleus may vary. The different nuclear types of an element are called isotopes. An isotope is designated by the name or chemical symbol for an element and the total number of protons and neutrons in the nucleus. Thus we have uranium-233 (U^{233}), uranium-234 (U^{234}), and so on.

Energy is released in three general types of nuclear reactions. The first of these is the process of radioactive decay, in which unstable nuclei achieve stable configurations through the release of gamma rays or subatomic particles. Radioactivity is the source of the heat generated in the crust and upper mantle of the earth. The second major type of nuclear reaction is the splitting of a large nucleus into a number of moderate-sized fragments accompanied by release of gamma rays and subatomic particles. This process is called fission. Finally, energy may be released in the fusion, or joining together, of small nuclei.

Although energy is released in a large number of nuclear reactions, for practical purposes, very few reactions yield net useful energy because more energy must be expended to achieve the conditions necessary for the reaction than is derived from the reaction itself. A notable exception is the fission of uranium. One isotope of uranium (U^{235}) can be used as a nuclear fuel because its nuclei can be split by the impact of slow neutrons (neutrons traveling perhaps a mile per second) and — what is most important — each fission produces an average of 2.5 additional neutrons. These neutrons can induce still more fissions, so that a self-sustaining (or chain) reaction is possible.

Some isotopes which are not readily fissionable can be converted, through absorption of a neutron, to ones which are. When a nucleus of uranium-238 absorbs a neutron it undergoes changes leading to the formation of an isotope of the manmade element plutonium (Pu^{239}). Plutonium is fissionable, reacting with slow neutrons much as U^{235} does. America's second atomic bomb was a plutonium bomb. Similarly, an isotope of thorium (Th^{232}) can be converted to fissionable U^{233} by absorption of a neutron. The process of producing a fissionable fuel from a nonfissionable one is called breeding, and the isotopes from which the fissionable fuel is derived are called fertile.

At first, nature seemed to conspire against any use of fission. The only fissionable fuel, U^{235}, was hopelessly diluted, 140 atoms to 1, by U^{238}. There was no way, then, to separate one isotope from the other. A fissionable fuel could be produced from U^{238}, but only neutrons from concentrated U^{235} could do the job. No wonder that, even with knowledge of uranium fission, no German bomb project got off the ground.

Some scientists suspected that only under certain conditions could U^{238} trap neutrons and extinguish the nuclear reaction. They found that slow neutrons could split U^{235} nuclei but were not absorbed by U^{238}. Enrico Fermi, the man who had split the atom without knowing it, supervised construction of the first atomic pile, or nuclear reactor, a project begun at Columbia University and culminated on December 2, 1942 in the squash court beneath the stands of the University of Chicago's Stagg Field. It was fashioned of a carefully calculated arrangement of graphite and uranium oxide bricks, interspersed with rods of cadmium and boron. Graphite was the moderator, slowing neutrons to speeds that would ensure their capture by U^{235} and not U^{238}. The cadmium and boron rods, avid neutron scavengers, could be inserted or withdrawn to control the rate of the reaction. Fermi turned on the world's first reactor at 3:25 in the afternoon. At 3:53 he turned it off, proving that a nuclear chain reaction could be started, maintained at a desired rate, and stopped at will. Today's reactors surpass Fermi's pile in both scale and sophistication of design, but their fundamental principles are the same.

Most of a nuclear power plant is basically the same as a fossil fuel plant. Only the primary source of energy differs. In both types of plant, heat from the energy source is usually used to produce steam, which turns a turbine, which drives a generator to set up a current of electricity. We will focus here on the initial step of the process, production of heat in the nuclear reactor.

Every reactor has a fuel supply, a means of controlling the rate of fission, and a means of transferring heat from the reactor to wherever it will be used. There are many variations in reactor design, but the most significant feature of a reactor is whether it is a net consumer or a net producer of fuel. Because 2.5 neutrons are produced in every fission of U^{235}, while only one is needed to maintain the reaction, and because neutrons can convert U^{238} and Th^{232} to fissionable fuel, it becomes possible for a reactor to produce up to 1.5 times as much fuel as it uses. By the ratio of fuel produced to fuel consumed, a reactor is classified as a burner, a converter, or a breeder.

Most of today's reactors are burners, able to release only 1 or 2 percent of the energy potentially available from uranium. The fuel for a burner is uranium dioxide which has been enriched to contain about 3 percent U^{235}. (The two isotopes of uranium can now be separated on the basis of the 1 percent difference in their weights.) The uranium is sealed into metal tubes which — when they are functioning properly — prevent the escape of radioactive fission products from the fuel rods. Ordinary water acts as both moderator and coolant. As in Fermi's reactor, movable rods of boron or cadmium regulate the rate of fission. Additional boron, as boric acid, may be dissolved in the moderator. Steam generated from the cooling water in the reactor may be used directly to drive a turbine, or heat from the coolant may be transferred through a heat exchanger to a separate system for steam production.

Reactors which create significant amounts of fuel, but not as much as they consume, are called converters. Their more favorable conversion rate is accomplished by modification of the fuel rods, which are blanketed in layers of fertile material. Conversion of Th^{232} to U^{233} proceeds readily using the slow neutrons produced in a conventional water-cooled, water-moderated reactor. Although present conversion rates are low, converters with ratios in the vicinity of 1 can probably be built to operate on the Th^{232} - U^{233} cycle.

The United States has given most attention to a converter that uses helium as a coolant, graphite as a moderator, and highly enriched uranium dicarbide fuel containing 90 percent U^{235}, with thorium dicarbide as the fertile material for breeding. Gas turbines replace the usual steam turbines, eliminating the steam cycle. These and possibly other types of converters will be used increasingly in the near future as a compromise between the present generation of wasteful burners and the breeders that we cannot yet build.

A breeder reactor produces more fuel than it uses. Breeders based on the Th^{232} - U^{233} cycle have been designed and show promise. The ratio of fuel production to consumption is expected to be relatively low, perhaps only 1.06 or 1.07, and for this reason most countries are placing their hopes on breeders that utilize fast neutrons and the plutonium cycle.

The fast breeder is fundamentally very different from reactors using slow neutrons. The moderator has been eliminated. In order to maintain the reaction, fissionable material must be packed in concentrated form in a small space. The fuel is surrounded by fertile material, which is U^{238} left over from the production of fuel elements. It is not possible to use water for a coolant because of its simultaneous role as moderator and because it is unable to transport heat fast enough from the reactor's intensely reactive core. Coolant for a fast breeder must have a high capacity for heat transfer and little or no tendency to interact with neutrons. The leading candidates for coolants are helium under pressure and liquid sodium. Although technical and economic obstacles prevent construction of a large fast breeder reactor, several small ones are already operating in the United States and Europe and several larger ones are scheduled for completion in Europe in the mid-1970s.

It is customary to speak of nuclear fuel reserves in terms of the price per pound of the uranium oxide U_3O_8. Known and estimated total supplies of uranium in various price ranges are shown in table 6-1. This table also presents estimates of uranium requirements between now and the year 2000, based on projections of a nuclear power industry depending on burner reactors. It is clear that reserves of inexpensive uranium will be exhausted by the end of the century if we continue to rely on reactors that consume more fuel than they produce. Even if conversion ratios are improved, so long as they are less than 1, the initial supply of U^{235} and all the fuel it breeds will be depleted in a relatively short time, leaving a useless

Table 6-1.   Estimated Reserves of Uranium in the U.S. as a Function of Price

Price per Pound of U_3O_8	Tons of Known or Suspected Uranium (as U_3O_8) at This Price
less than $8	594,000
$8 to $10	346,000
$10 to $15	510,000
$15 to $30	790,000
$30 to $50	7,760,000
$50 to $100	15,000,000

These estimates are likely to be low, as estimates of total reserves often are. Uranium prospecting was vigorous in the past, but it has slowed down in recent years because the growth of the nuclear power industry has not been hampered by shortages or predicted shortages of fuel. Cumulative consumption of U^3O^8 may reach 1,600,000 tons by the year 2000. According to the table above, the price of uranium will have risen from less than $8 per pound to about $18 per pound by that time. This increase, while it seems large, would have a negligible effect on the cost of electricity.

From Atomic Energy Commission, *Potential Nuclear Growth Patterns*, (1970), as quoted in H. Hottel and J. Howard, *New Energy Technology* (Cambridge, Mass.: MIT Press, 1971), p. 34.

residue of fertile material that cannot be bred because there is nothing to breed it with. For this reason the fast breeder program in the United States has been speeded up almost to the proportions of a crash project.

But economic statistics are misleading. The price of fuel for a nuclear power plant is a relatively small cost, in contrast to overall costs for nuclear energy and fuel costs for fossil plants. If the cost of uranium were multiplied by 70, the cost of electricity from a nuclear plant would barely double. Another look at table 6-1 will indicate that neither the amount nor the cost of fuel will limit us, for there is plenty of uranium. However, social and environmental penalties for mining it may preclude its use. For this reason, some experts believe that, unless fast breeders become a reality before high grade ores of U^{235} are depleted, the history of nuclear energy will be very brief; whereas if fast breeder technology is perfected in time, there will be enough uranium and thorium ore to last for thousands of years.[2]

The first full-scale commercial nuclear power plant began operating in the U.S.S.R. in 1954. The first nuclear plant in the United States was placed in service in Shippingport, Pennsylvania, in 1957. It was followed by others in Illinois (1960), Massachusetts (1961), New York (1962), and California (1963). This was a period of acknowledged experiment and testing. The plants were relatively small, each producing as much electricity as a small fossil fuel plant. The largest of the five had a capacity of 265 megawatts. But orders for nuclear power plants began to pour in, and by the end of 1970, nineteen operating plants represented 2 percent of America's electric utility capacity, while fifty-three additional plants were under construction and thirty-four were being planned. By mid-1971 there were twenty-five plants in operation, forty-eight under construction, and sixty-three on order. The Federal Power Commission estimates that the percentage of the total electric generating capacity contributed by nuclear plants will grow from 2 percent in 1970 to 24 percent in 1980 and 41 percent in 1990.[3] The first large-scale commercial breeders are expected to begin operation in the mid-1980s. Meanwhile, the size of individual nuclear units will increase from approximately 1000 megawatts in the early 1970s to between 3000 and 3500 megawatts, with several of these units comprising a nuclear power facility.

General environmental concern and local action to block construction, coupled with unexpected cost increases and technical problems, have recently slowed the growth of nuclear power. In the future, these factors will probably be offset by increasing shortages of fossil fuels and rigid air quality standards. Nuclear energy today is heavily subsidized by fossil fuels at all stages from the mining of uranium to the actual generation of electricity. (Because nuclear plants must operate at lower temperatures than fossil plants for reasons of safety, some nuclear plants use a conventional fossil fuel plant as a final step to boost the temperature of the

Above, the external appearance of this nuclear power plant on the shore of Lake Michigan is like that of any other business establishment. *Below,* a clue to its complexity may be found in its control room. (Photos courtesy of Marine Studies Center, University of Wisconsin.)

steam.) Obviously, without fossil fuels, we would not have nuclear power either.

Nuclear power plants have certain advantages over fossil fuel plants. Outstanding among these is their lack of kinds of air pollution associated with burning conventional fuels. Nuclear fuel is many thousands of times less bulky than coal, and thus unsightly storage problems and endless processions of coal cars to and from the power plant are eliminated. Because the amount of fuel required is small and partial refueling is done only once or twice a year, transportation costs are small, making the cost of a nuclear plant practically independent of its location. Thus nuclear energy may be a good choice for areas far from a source of fossil fuel. Nuclear plants are quiet. They are less ugly than fossil fuel plants. Enthusiasts boast that although there have been accidents, the impact has never been felt beyond the confines of the power plant. But skeptics warn that next time the public may not be so lucky.

Optimists describe deserts bursting into bloom, watered by fresh water from the sea, and abundant food and comfortable lives for all — these miracles to be mediated by cheap, abundant nuclear energy. But there are serious fallacies behind these rosy pictures. Because electricity from nuclear power plants is very expensive on a small scale, it is not an alternative for developing countries with small needs for power. Proposed nuclear agro-industrial complexes would provide some relief, but not much. Plans call for large-scale desalination of seawater (for irrigation of surrounding intensively farmed areas) to be coupled with production of fertilizer and other industries. The food produced at such an establishment would feed about 3 million people. But twenty-three such plants would be required each year, at a cost of 41 billion dollars, just to keep pace with population growth.[4] It looks as if scale, time, and cost are working against us. In addition, major technological successes, including modern agriculture, tend to be major causes of environmental problems.

There are uses for the fissionable atom besides generating electricity. All of them involve some mix of the problems already discussed. Our preoccupation with nuclear weapons continues. The first nuclear-powered submarines have won many advocates for a nuclear navy, with its practical and strategical advantages. The Atomic Energy Commission seems eager to apply its expertise to a variety of peaceful pursuits as well. These include nuclear blasts to assist mining operations, to enlarge existing harbors or create new ones, and to facilitate the production of natural gas from impermeable formations. Considerable effort has been invested in attempts to increase gas production. The hazard of radioactivity remains, although it has been reduced, and possible seismic effects must always be considered. In 1971, the Chairman of the Federal Power Commission told a Congressional committee that an analysis by his staff indicated it would be twenty years before gas produced from such "stimulated" reservoirs would approximate 10 percent of the gas consumed in 1969. This level of production would be brought about by four thousand nuclear explosions of 100 kilotons each in one thousand wells, during the next twenty years.

At the other end of the scale, there are a number of experimental dogs whose hearts beat at the command of plutonium-powered pacemakers. The variety of potential nuclear-powered gadgets is limited only by the scope of man's imagination.

Nevertheless, nuclear energy is failing by a wide margin to fulfill the predictions of the 1950s, when the chairman of the Atomic Energy Commission thought it not unlikely that "our children will enjoy in their homes electrical power too cheap to meter, will know of great periodic regional famines in the world only as matters of history . . . and will experience a life span far longer than ours." Whether even the peaceful atom can be exploited without excessively high social costs is by no means certain.

The Atomic Energy Commission was established after World War II as a civilian agency that would search out, promote, and supervise peaceful uses of the atom. After a quarter of a century, about seven thousand full-time employees and many more workers under grants and contracts still pursue the original mission, at a cost of around 2 billion dollars a year to the American taxpayer. Many portentous questions remain unanswered. Yet we are embarking on a course that will make us dependent on nuclear energy by the end of the century.

The problems related to the use of nuclear energy fall into three classes. Most important of these, imprinted in our minds by the victims of Hiroshima and Nagasaki and by worldwide fallout from testing of nuclear weapons, is radiation. There are also the social and ecological consequences of mining large amounts of low grade uranium and thorium ore, and the biological and climatic effects of producing enormous amounts of waste heat. On a second look, the revolution to be brought about by nuclear energy appears to be merely an extension of well-known patterns and problems.

It is important to understand something of the nature and measurement of radiation and how it affects matter and living things. Radiation released during nuclear fissions or transformations of radioactive nuclei consists of subatomic particles (positive and negative electrons, alpha particles, and neutrons) and gamma rays. This radiation is measured in rads or millirads. (There are one thousand millirads in one rad.) A rad is simply the measure of the amount of energy radiation imparts to matter. Technically, it corresponds to absorption of 100 ergs of energy per gram of matter, but all that need concern us here is the meaning of the term rad and the fact that 100 ergs is not very much. Another measure of radiation is the rem, which is generally equivalent to a rad except in the case of very energetic alpha radiation, when a rem may be ten or twenty times as large as a rad.

There are more than a thousand different radioactive isotopes, many of them manmade. With each isotope is associated a characteristic time called the half-life, a statistical property giving the time required for half the nuclei to decay. After one half-life has elapsed, half of the original nuclei remain; after two half-lives, one quarter; after three half-lives, one eighth; and so on. The half-life varies tremendously from isotope to isotope. Half of the nuclei of polonium-212 decay in less than a millionth of a second, for example, while the half-life of thorium-232 is 10 billion years.

Particles emitted from radioactive nuclei usually carry more than a million times the energy of the same particles at rest. They lose this energy bit by bit in collisions with atoms, which may become electrically charged, or "ionized," as a result. Ions have different chemical properties from the atoms or molecules that gave rise to them. If enough ionizations occur, the function of an entire cell can be disrupted. This is how radiation damages living things.

The distance a charged particle travels before coming to rest is related to its size. The heavier the particle, the shorter its disruptive path. If they have the same amount of energy to begin with, either an alpha particle or an electron will cause the same number of ionizations (typically about 100,000) before coming to rest. The heavy alpha particle, however, travels only one-sixtieth as far as the electron. Thus its path, while shortest, is the most densely devastated. Alpha particles are extremely dangerous when they arise from a source within the body — in the bones or lungs, for example — but virtually harmless if created only an inch outside of the body. Exposure of the whole body to a concentrated dose of more than 600 rads brings death in a few weeks, from irreparable damage to bone marrow, lymph nodes, and spleen. A dose this size causes an average of more than a million ionizations in each cell of the body.[5] With an acute dose of 100 to 600 rads, the same types of changes occur although the chances of recovery are increased as the dose becomes lower. Doses of 100 rads or more would be experienced only in nuclear war, in a major nuclear accident, or in radiation therapy.

The causal relationship between radiation and cancer and leukemia is well documented although incompletely understood. Cancers have been produced in adults by single or extended doses of 100 rads. Doses to the fetus of only 5 rads have been implicated in the development of cancer. There was an increased incidence of leukemia among early radiologists and among survivors of Hiroshima and Nagasaki. Workers painting radium dials in the years before World War II moistened and shaped their brushes on their tongues. The radioactive material they ingested in this way collected in their bones, where it caused a high incidence of cancer. Deposits of radioactive iodine in the thyroid can cause cancer of this organ. Inhalation of radon and its radioactive "daughters" (successive stages in the disintegration of uranium) has led to a high rate of lung cancer among uranium miners.

Chronic exposure to radiation also results in a shortening of the lifespan which cannot be attributed to any particular cause of death. It has been compared to the aging process, and is probably due to cumulative minor damage to cells in all parts of the body.

If radiation causes ionizations in the hereditary material (DNA) of a sperm or an egg cell, a change or mutation can occur in the genetic information passed on by parent to offspring. Mutations are likely to cause abnormality or death of organisms developing from affected cells. The defects are inheritable and may be passed on to future generations. Not all genetic damage is as dramatic as a stillbirth, abortion, or gross defect. Many mutations are subtle and at present impossible to detect. They may predispose an organism to certain diseases, or render it generally less fit than one not carrying the mutation.

Some of the radiation to which we are exposed is natural, arising from cosmic rays, solar radiation, and radioactive elements in the earth. It varies between 80 and 150 millirads per person per year. It varies both with the latitude and the altitude at which a person lives and with the composition of the rocks beneath him and the house around him. Because man has lived in this level of radiation for about 2 million years, many people consider it safe and conclude that additional amounts of radiation of the same order of magnitude are also safe — or at least, insignificant. However, it is reasonable to assume (even though we cannot measure it) that some fraction of the cancers, genetic defects, and diseases of aging from which man has always suffered is caused by the natural radiation around and in him, and that any increase in exposure will bring about a corresponding increase in radiation-induced disabilities.

Exposure to radiation in medical diagnosis and therapy adds, on the average, 65 millirads per year per person. Although exposure from a single diagnostic X ray has been reduced through improved equipment and technique, overall medical exposure is increasing because of increased reliance on radiological methods.

People do not become disturbed about natural background radiation or medical exposure. The first is unavoidable and the second is a matter of personal choice, in which the benefits are usually clearly seen to outweigh the hazards. But when people are told they will be exposed to manmade radiation without being given a chance to say no, they become alarmed and angry. Two different curves describe the acceptable balance between benefit and risk, depending on whether the activity is voluntary or involuntary.[6] In general, when the decision is his own, a person is willing to accept between 1000 and 10,000 times the risk that he will tolerate if the choice is made for him. Thus a skydiver may rage at the thought of having fluoride added to his drinking water.

Much of the controversy about radiation standards for the general population and for workers who receive occupational exposure centers around the existence or nonexistence of a threshold dose below which no damage occurs. For a few effects, such as the development of cataracts, a fairly definite threshold has been established. Certain types of damage occur only when the radiation has more than some minimum amount of energy. Factors like nutritional status and oxygen concentration influence the extent of damage. The body has mechanisms, which we are only beginning to understand, by which it repairs radiation damage. An acute dose overwhelms the capacity for repair, and this may explain why a given dose becomes less harmful as it is spread out over time. It also helps to explain why cells that are growing, dividing, or metabolizing rapidly are more sensitive than inactive cells, and why a fetus is more sensitive than an adult. The active, growing, dividing cell does not have time to repair itself before the damage is expressed.

Chronic, low level exposure concerns us most, and for this, data become scarce and inconclusive. Many radiologists believe that for cancer induction there may be a practical threshold if not an absolute one. A latent period occurs between the time of exposure and the development of a cancer. This latent period increases as the dose rate and the dose decrease. When it becomes longer than a normal lifespan, you will die of something else before your cancer develops.

For genetic effects, no such practical threshold can exist. Although statistics for people are lacking, experiments with lower animals suggest that there is no threshold for genetic effects, although the effects do depend on the dose rate.

Taking into account all that is known about the effects of radiation on man, the Federal Radiation Council and the National Council on Radiation Protection propose radiation standards to the Environmental Protection Agency and the Atomic Energy Commission, which are empowered to establish and enforce them. The Atomic Energy Commission sets industrial standards governing occupational doses, doses to the general public, and maximum emission levels for various types of facilities. The Environmental Protection Agency is responsible for environmental quality standards. In practice, the real work of regulation is carried out by the Atomic Energy Commission, as it has always been.

Standards for doses of manmade radiation have been revised repeatedly since the first occupational limits were established in England more than fifty years ago. In the United States they are currently set at an average of 0.17 rem per person per year for the general population, with a maximum of 0.5 rem per year for an individual. Radiation workers are limited to 5 rem per year. Medical exposure is excepted from these limits. Although the permissible doses of manmade radiation approximate those from natural sources, there is widespread concern that the standards are not strict enough. But every activity bears some risk. If three thousand extra cancer deaths each year seem unthinkable, remember that many people die whenever London or New York has a good smog. While this is no excuse for killing people with radiation, it may put things into better perspective. A cost-benefit analysis might show the annual per capita medical cost for nuclear energy to be many times less than the cost of preventing this expense.

Regulating the release of manmade radioactivity could reach great levels of complexity. It is not enough to monitor the release of each individual radioisotope. Each one should be followed through all the physical, chemical, and biological pathways by which it is concentrated or dispersed. Because each nuclear facility is a unique case, each facility, in the extreme, would require its own standards, in order to insure a uniform and acceptable level of risk. This would make regulation the problem rather than the solution.

It is misleading to compare the total exposure to manmade radiation with the exposure to natural radiation because the radioisotopes involved are not the same. Some radioisotopes released by man do not occur in nature at all. The dangerous iodine-131 is one of these. Plutonium-239 is another. It is not as if we walk around in some level of generalized radiation which is suddenly doubled. We are exposing ourselves to brand new hazards. A few examples should illustrate the problems faced by makers of standards. Radioactive krypton, a chemically inert gas, is considered practically harmless, and has been intentionally vented to the atmosphere. Similar amounts of iodine-131, zeroing in on thyroid glands, would be intolerable. The rate at which radioisotopes are released is only one factor related to their ultimate concentration in the environment. For gaseous emissions, prevailing wind and precipitation patterns must be considered. For isotopes in liquid effluents, the characteristics of the body of water receiving them are important. Radioactivity from one nuclear power plant using the cooling capacity of a swiftly flowing river might have no significant effect on the environment. But what about the second plant on the same river? Or the tenth?

The situation with lakes is even more complex. Lakes are fed by some rivers and drained by others. Each lake has its characteristic flushing time, which is the time required for a complete renewal of its water. Despite the wailing over Lake Erie, this lake could be restored, not necessarily to its former condition, but to a satisfactory condition, within a short time, because its flushing time is about three years. Lake Michigan, on the other hand, has a flushing time of many decades — perhaps as much as a century; and the flushing time for Lake Superior is five hundred to one thousand years. While radioisotopes discharged into Lake Erie are soon washed out again, those discharged into Lake Michigan or Lake Superior can accumulate to dangerous levels.

Above, an aerial view of an experiment on the effects of long-term chronic exposure to radiation. The trees in the center died after being exposed to 20 hours of gamma radiation per day for 6 months. *Below,* a closer view of the damage to trees and vegetation in the same experiment. (Photos courtesy of Brookhaven National Laboratory.)

If mixing of the water is incomplete, radioisotopes may collect in certain regions of the lake. Concentration of radioisotopes in the food chain is a matter of major concern. Research into these areas is clearly indicated.

Meanwhile, simple standards have been set following recommendations of national and international advisory bodies. The standards accept the most conservative interpretation of available data and incorporate an additional safety factor. Necessary measurements can be made, so that the standards can be enforced. This is important. As we should have learned from Prohibition and attempts to legislate sexual activities, an unenforced or unenforceable law can be worse than none at all.

Until now, standards have been set with only man's safety in mind. What of the effects of radiation on other species? When we think of protecting man, we mean protecting individual men, not *Homo sapiens* in general. With other species, however, the concern is for survival of populations, the fate of an individual mattering very little. In many species, individuals are damaged but populations adapt and survive. The implications of this adaptability are profound. Chronic exposure to radioactivity will probably increase the mutation rate in most or all organisms, leading to increased variation and increased possibilities for adaptation. Think of the consequences for organisms like bacteria, insects, and rats, creatures with short life cycles which are already adapting faster than we can devise ways to control them.

In considering hazards of radiation, it is easy to focus on the power plant and forget the other steps in the total nuclear energy cycle. The problems begin in the uranium mines, where miners in the past have been exposed to dangerous levels of radon and its daughters. During the last fifteen years the plight of uranium miners has received considerable attention, with the result that the radiation hazard has been greatly reduced. Miners now contracting lung cancer were exposed to levels of radiation up to twenty times the maximum permitted today. Unfortunately, it will be at least another decade before we have enough data to show whether today's standards are adequate.

Another problem is the radioactive residue left from treating uranium ores. A typical mill must dispose of about 10 curies* of radium each day. There is no satisfactory way to deal with the millions of tons of radium-rich wastes which lie exposed and vulnerable to erosion by wind and water. Part of this material was once used as landfill and homes were built on it. But that no longer seems like a very good idea.

Processing uranium fuel involves a number of steps. These are primarily chemical and metallurgical processes with which we have had a great deal of experience. The hazards are negligible. After being milled

---

*A curie is a quantity of a radioactive isotope equivalent in activity to 1 gram of radium, in which there are 37 billion nuclear disintegrations per second.

into "yellowcake," ore is refined into "orange oxide," then hydrogenated and converted to "green salt." The green salt is converted to uranium hexafluoride for enrichment at one of the three gaseous diffusion plants located at Oak Ridge, Paducah, and Portsmouth. The enrichment step, in which U^{235} is concentrated severalfold, requires an enormous amount of energy, which is derived from burning high sulfur coal. The gaseous diffusion plants use more than 45 billion kilowatt-hours of electricity annually, more than is generated in the power plants for which they provide nuclear fuel. (Not all of the enriched uranium is destined for the generation of electricity, but much of it is.)

The greatest dangers in the entire uranium fuel cycle arise during reprocessing of used fuel elements, owing to the variety of concentrated radioactive materials that must be handled without the built-in protective mechanisms that exist in the power plant itself. Unspent uranium and newly created plutonium are separated for use in new fuel elements; radioactive fission products and byproducts are separated for disposal or storage. Standards for chemical reprocessing plants are more lenient than those for power plants, an admission that the radioactivity cannot be adequately contained. The advent of a successful fast breeder will increase the problem because of the quantities of plutonium that must be processed. Plutonium is the deadliest element known. One millionth of a gram injected into the bloodstream of a dog can cause bone cancer.

Transportation of nuclear fuels and radioactive wastes poses another potential hazard, in the event of accident or sabotage. So far the record has been good. But as traffic in radioactive materials increases, accidents are bound to occur. If plutonium becomes a major energy source worldwide, let us hope for more political stability and individual sanity than exist today. There is little comfort in knowing that nuclear weapons so far have kept an uneasy peace among nations possessing them, when anyone with a chunk of plutonium can build a bomb.

Although the probability of a major accident in a nuclear power plant is very very small, it can never be zero. The consequences of such an accident can be as great as you care to imagine. How small is very small? One accident in a thousand years of a reactor's operation? With 100 reactors, then, there would be a major accident every ten years. With 1000 reactors, there would be a disaster each year. We do not yet have enough experience to know how small a very small probability is. Placing the reactor underground wherever geology permits offers an additional measure of real security as well as a psychological advantage.

Nuclear power plants continuously produce gaseous, liquid, and solid radioactive wastes. A typical 800 megawatt plant produces the same number and kinds of nuclear fissions in ten years' operation as were produced in all the aboveground testing of nuclear weapons. More than 3 million curies of strontium-90, for example, are formed each year in such a plant. In addition to the mass of radioactive fission products which is very nearly equal to the mass of fuel consumed, the materials of the reactor itself and the coolant are made radioactive by neutron bombardment. Although the idea of this much radioactivity is frightening, most reactors release substantially less radioactivity to the environment than the regulations allow. There is no immediate cause for alarm. The problem, again, is growth. Safely within the limits now, we will approach the danger point rapidly as exponential growth of nuclear power continues.

By committing ourselves to the use of nuclear fission we are also committing ourselves and our descendants to the permanent custodianship of staggering amounts of radioactive wastes. Already, hundreds of millions of curies are awaiting development of a satisfactory disposal scheme. The plan is to concentrate the wastes as much as possible, solidify them in an insoluble form, and bury them in special containers deep within geologically stable structures safe from large earth movements, invasion of groundwater, and incursions by man. Certain salt beds meet the criteria for burial sites, as do some water-free granite structures near the oceans. No one has a plan for what to do if, despite all precautions, something goes wrong at the graveyard, although theoretically the wastes can be retrieved. None other than the Director of the Oak Ridge National Laboratory has compared the social commitment to nuclear energy with the social commitment made thousands of years ago to agriculture. It is a turning point. There can be no going back.

By the time it reaches the power plant, the concentrated energy in uranium fuel gives uranium a tremendous advantage over coal, many tons of which must be burned each day. Much of this advantage is negated in the mines, however, where a ton of earth must be dug up to get a pound of uranium oxide. And very soon these relatively rich deposits will be gone.

More than half the uranium ore in this country is amenable to mining with open pit methods, which resemble the strip mining of coal. To the usual environmental problems is added radioactivity. In addition, the volume of waste remaining after uranium has been removed is at least as great as the amount of material that was dug up. The earth has been grossly disturbed. Reclamation is at best difficult and at worst impossible.

The inexorable laws of the universe decree that all the energy we use will ultimately be degraded to useless heat. The production of electricity in the present generation of nuclear power plants is a relatively inefficient process because, for reasons of safety, the nuclear plant must operate at a lower temperature than the fossil fuel plant. The efficiency of the steam cycle depends on the initial and final temperatures of the system, with the greatest efficiency resulting from the largest temperature differential. Future generations of nuclear reactors will probably overcome this problem, making nuclear power plants no less efficient than fossil fuel plants. But any steam plant discharges large amounts of heat to the environment, whether to the atmosphere or to cooling water. When you put heat somewhere, something happens. Most of the things that happen are considered undesirable by most people, although some waste heat could be used to our advantage. The effects and possible uses of waste heat will be considered in chapter 8.

While we are already living with some of the costs and some of the benefits of energy derived from splitting large atomic nuclei, the reverse process, energy from the fusion or joining of small nuclei, remains only an enticing possibility. Scientists have been observing fusions of light nuclei in the laboratory for several decades. Their studies have given insights into the nature of matter, but because of the tremendous amount of energy required to induce fusion and the infrequency of the fusions themselves, they have failed to advance us significantly toward the goal of gaining useful energy from the fusion process. Since about 1950, several nations have pursued research programs aimed specifically at making fusion an energy source for man.

Although many fusion reactions occur in the stars, earthlings are limited to those involving the two heavy isotopes of hydrogen, deuterium and tritium. Tritium is so rare in nature that it must be made artificially if we are to obtain useful quantities of it. The best way to make it is by bombarding the lithium isotope of mass 6 with neutrons.

Technology for separation of deuterium from sea water is well developed. It poses no foreseeable environmental problems even if conducted on a large scale. If we were to remove only 1 percent of the deuterium from the sea, we would have fuel equivalent in energy value to half a million times the earth's initial stock of fossil fuels.

The situation with lithium is quite different. The isotope of interest, Li^6, comprises only 7 percent of natural lithium. Known and inferred reserves of lithium are small because this metal, unlike most, does not tend to concentrate in ores, but occurs diffusely throughout the oceans and the earth's crust. Geologists believe that the minable lithium deposits of the world would yield only about as much energy as the world's initial supply of fossil fuels. This may or may not be a cause for pessimism. If we actually develop a fusion reactor based on deuterium and tritium fuel, the major problems of fusion will have been solved. Use of the deuterium-deuterium reaction would be only one technical step away. The question of recovering lithium from seawater is controversial. Some forecasters look to the sea as a source of most of our future minerals, but others point out that mining the sea will probably be incompatible with farming it, which we also propose to do.

Controlled nuclear fusion looks like a nearly ideal source of energy (although coal did, too, a couple of centuries ago, as did nuclear fission in the 1950s). There appear to be no problems in obtaining cheap and abundant fuel if deuterium is used. The fusion plant itself would provide both heat and electricity with high efficiency (perhaps 80 or 90 percent) and few unwanted side effects. The main end product would be helium, for which there are many uses. The amount of radioactive waste would be thousands of times smaller than for equivalent fission plants. The only radioactive material formed in large amounts would be tritium, for which improved methods of handling are needed.

It is difficult to talk about the safety of a fusion reactor since no one is sure what one will be like. However, while there is always the possibility of fission reactions getting out of hand, no such chance exists with fusion. It is hard enough to get sustained fusion started in the first place. Anything that went wrong would quickly quench the reaction.

Although fusion reactors could be the permanent solution to world energy problems, the economic and technical problems may be insurmountable. Why is it such an immense step from an uncontrolled fusion reaction (an H-bomb) to a controlled one, when the step from fission bomb to reactor is so easy? Fission can occur any time that a fissionable nucleus captures a neutron. You need only to make sure that the right number of neutrons is available to an appropriate number of fissionable nuclei. Fusion, on the other hand, occurs only in the stars or under simulated stellar conditions. While nuclear fusion is the source of the heat of the stars, only the heat of the stars allows fusion to proceed. Fusion requires that two nuclei come together with enough energy to overcome the repulsive forces that normally exist between particles of like charge, enabling the very powerful intranuclear forces which act only at exceedingly short range to come into play.

At temperatures of tens of millions of degrees, atoms as we know them cannot exist. Instead, matter is a plasma of free nuclei and free electrons. Even in this state, fusions are relatively infrequent until temperatures of some 100 million degrees are reached. At 100 million degrees, the reaction between deuterium and tritium sustains itself at a rate sufficient to yield useful energy. Even higher temperatures are required for the deuterium-deuterium reaction to be of interest, and for this reason scientists are concentrating on the deuterium-tritium reaction even though its usefulness will be limited by availability of lithium.

Three conditions must be achieved simultaneously in a fusion reactor: a temperature of at least 100 million degrees, a plasma dense enough to insure frequent collisions of nuclei (by most standards, this would be a fairly good vacuum), and a confinement time long enough for the process to yield net energy. What container do you use for a plasma heated to 100 million degrees? No material container would do the job. On a small scale, for a short time, and at great cost in energy, plasmas have been confined within strategic arrangements of intense magnetic fields, called magnetic bottles. No type of magnetic bottle has yet met all the requirements. Some leak too much. Some require so much energy for their creation and maintenance that it is doubtful whether they would ever yield net energy.

Even with an adequate confinement system, the structural materials of the reactor would be subjected to neutron bombardment several times as intense as that inflicted on today's fission reactors. Since one of the major problems with fission is maintaining the structural integrity of the reactor under the onslaught of high energy neutrons, some engineers predict that the useful life of a fusion reactor may be impractically short. To combat this difficulty, some proposed reactor designs rely heavily on unusual materials. One preliminary design, for example, calls for large quantities of the rare metal niobium. Total quantities of this resource appear prohibitively small. Whether the leaking neutrons damage the reactor or not, they will make it radioactive. These radioactive materials constitute a disposal problem. How serious this problem could become depends on both the ultimate design of fusion reactors and their useful lifetime.

Another means of releasing energy from fusion has met with limited success in the laboratory. When tiny pellets of frozen deuterium-tritium fuel are subjected to an intense laser pulse for a tenth of a billionth of a second, they explode like miniature hydrogen bombs. There is no need for a fancy magnetic bottle; no need, in fact, for a plasma. The problems are ones of scale. The fact that scientists can explode a few fusion grenades in the laboratory is no guarantee that a safe, economical — or even functional — 1000 megawatt power plant can be built on the same principles.

References

1. W. Laurence, *Men and Atoms — The Discovery, the Uses and the Future of Atomic Energy* (New York: Simon and Schuster, 1962).
2. M. K. Hubbert, *"Energy Resources"* in *Resources and Man*, National Academy of Sciences-National Research Council (San Francisco: W. H. Freeman and Company, 1969), pp. 219-228.
3. U.S., Federal Power Commission, *The 1970 National Power Survey* (Washington, D.C.: 1971).
4. P. Ehrlich and A. Ehrlich, *Population, Resources, Environment* (San Francisco: W. H. Freeman and Company, 1970), pp. 95-96.
5. J. Harte, R. Socolow, and J. Ginocchio, "Radiation," in *Patient Earth*, ed. J. Harte and R. Socolow (New York: Holt, Rinehart, and Winston, Inc., 1971), pp. 299-320
6. C. Starr, "Social Benefit Versus Technological Risk," *Science* 165 (1969): 1232-1238.

7

OTHER ENERGY SOURCES

It is the sun that shares our works.
The moon shares nothing. It is a sea.

When shall I say of the sun,
It is a sea; it shares nothing;

The sun no longer shares our works
And the earth is alive with creeping men,

Mechanical beetles never quite warm?

Wallace Stevens

Actual and predicted shortages of fossil fuels, doubts about full-scale commitment to nuclear fission, and the uncertainty of fusion have led to an inventory of all energy sources. For the long term, there are but three major sources of energy: nuclear fission (as accomplished in the breeder reactor), nuclear fusion, and the sun. The promise of these technologies is in the future, however, and promises are not always kept. Meanwhile, there are other sources of energy that can buy time or be of local importance — or even of general importance, should other technologies fall short of their goals.

It is illusory to hope that exploitation of energy can be without environmental effect, because by definition transfers and transformations of energy make things happen. However, the sources of energy discussed in this chapter have some significant advantages over fossil fuels and nuclear energy. Some of them represent the energy of continuing natural processes that cannot be depleted. For the most part they do not spew poisons into the air and water. There is no problem of radioactivity. And they require no mining, with its tremendous cost to people and the earth.

Wood was man's first and, for many thousands of years, his only fuel. Consumption of wood reached its peak in the United States around 1870, when it yielded about 840 billion kilowatt-hours of energy. But the relative importance of wood was already declining. In 1850, wood furnished 90 percent of the energy derived from all fuel and water power. Twenty years later it was only 73 percent. Today, we still burn about a quarter as much fuel wood as we did a century ago, but it supplies a minuscule fraction of the energy we use.

Those who glorify a simple way of life epitomized by cutting a supply of fuel wood for the winter may be solving their personal problems, but they are not solving the problems of the nation or of the world. In the future, wood will serve as only a most minor source of energy. Like fossil fuels, it may have such value as a raw material that it will become too precious to burn at all.

Other organic materials may also be burned as fuel. Whenever man has abandoned nomadic ways, he has been confronted with problems of waste disposal. One solution has been to burn human and agricultural wastes as fuel. This practice is widespread in poor countries and is still found in parts of the United States. It is not an unmitigated good. Air pollution was not invented during the Industrial Revolution in Britain.

Improved treatment of wastes is increasing their value as fuel at the same time that other technologies are beginning to compete for them. Bagasse, a fibrous by-product of the sugar cane industry and an important non-commercial fuel in some areas, is now in demand for the manufacture of paper. Other agricultural wastes which once were burned are being converted to fertilizer, feed, and food supplements. If we begin to make use of the tons of animal wastes that now pollute the air and water, we may have to decide whether they should be returned to the earth as fertilizer or converted to methane by anaerobic digestion. It has been calculated that conversion to methane of all animal wastes generated in the United States by a process under study would supply methane at a rate equal to about half our current rate of natural gas consumption.[1]

In Europe, where supplies of fuel have long been limited, a large percentage of municipal solid waste is burned to generate electricity and provide heat. Government agencies in the United States, along with universities and electric utilities, are beginning to examine new processes for converting solid waste to fuel. The most promising of these would convert the garbage to pipeline quality gas or liquid fuel. Two processes, both at the pilot plant stage, yield low-sulfur fuels with 50 to 75 percent of the energy value of the original material. One enthusiast estimates that the efficient conversion of garbage to power could furnish up to 6 percent of the energy needs of the United States at the same time that it solved a multi-billion dollar disposal problem.[1] Of course there are difficulties. One of them is poisonous combustion gases from plastics, which might pose a pollution problem. And the more garbage we recycle, the less there will be to burn.

Competing uses for garbage are not yet a problem — today, the problem is that nobody wants it. But whatever its other merits or demerits may be, human garbage is renewable; and one way or another, we will come to use it.

Use of wind power for transportation probably dates back to Stone Age times. Early man caught the wind in sails and moved himself over the water. He caught it on propellers and turned machines to do his work. The Dutch have a saying that God made other countries, but the Dutch [and

The bark *Swallow,* on her last whaling voyage from New Bedford in 1897. (Photo courtesy of the Whaling Museum, New Bedford, Mass.)

the windmill] made Holland. Western Europe's conquest of the globe and much of her early industry were powered by the wind. By the end of the nineteenth century, several hundred thousand windmills dotted rural and not-so-rural landscapes all over the world. At least thirty thousand of them were in Denmark, northern Germany, the Netherlands, and England. In the United States, they meant life itself in dry areas west of the Mississippi River.[2]

Now sailboats are all but extinct, except as pleasure craft. Many of the windmills have vanished; most of the rest stand idle. Even Holland, finding herself sitting on a bubble of natural gas, has turned her back on wind power. The number of windmill manufacturers in the United States dropped from more than twenty in the 1920s to two in 1972, reflecting the march of electricity into rural areas. Now the wind is dismissed as being insufficient to contribute significantly to the world's energy needs.

But a renewed interest in wind power among many nations belies that attitude, as do·the thousands of windmills manufactured in Australia, western Europe, and the U.S.S.R. for distribution around the world. As a truly renewable, non-depletable, and non-polluting energy source, the wind deserves respectful consideration.

The energy of the wind originates in the sun. The amount of wind power available for windmills has been estimated to be about 20 billion kilowatts, roughly three hundred times the amount of useful tidal power, six times the world's installed electric power, and 50 percent more than the total potential hydropower.[3]

The power in a current of air varies with the cube of its velocity. However, only 59 percent of this power is theoretically recoverable. A windmill works because some of the wind passes by. The efficiencies of modern windmills range between 50 and 80 percent. For an efficiency of 70 percent, if the wind's speed were 22 miles per hour, the power would be only 25 watts for each square foot swept by the propeller. The annual energy output for windmills varies between 10 kilowatt-hours in relatively calm areas and 50 kilowatt-hours in windy areas, for each square foot swept. The efficiency varies with the wind's velocity, but not linearly. Below a minimum speed, no useful power is generated; and when the speed surpasses a maximum value there is no corresponding increase in power. Since wind power depends on the cube of wind velocity, doubling the velocity produces an eightfold increase in power. One would think that a mill designed to function at normal wind velocities would be torn apart in a gale. But modern design insures that this will not happen, just as modern materials enable windmills to function reliably in climates as unfriendly as that of Antarctica.

The technology of windmills is well developed, as is the aerodynamic theory on which it is based. Over the centuries, almost every conceivable design has been tested. Each culture has developed models and materials characteristically its own, from the bamboo-and-sail structures of China to the picturesque Dutch mill to the spider-like wind pump that stands as a monument to the past on countless American farms. Modern materials such as plastics reinforced with glass have improved the performance of windmills and facilitated their manufacture. Extensive wind data are now available for many areas to assist in selecting sites.

The largest mills, ranging in capacity from 100 to 1000 kilowatts, are used for producing electricity. A number of them are hooked into Denmark's electric power grid and they function competitively with other types of power plants. Japan, another nation with a chronic fuel problem, is testing some large windmills.

Smaller mills with an output of 5 to 20 kilowatts are used to pump water or produce electricity in many isolated rural communities. This power is ample for refrigeration and illumination and the operation of radios. Small wind pumps, supplying less than 5 kilowatts, are usually used only for pumping water.

The disadvantages of wind power — its intermittent and unreliable nature, geographical limitations, and inability to supply large amounts of power for heavy industry — are well known. The only way to be sure of having power is to have a backup source ready for when the wind fails; and if this is so, why use the wind at all? For periods less than half an hour or an hour, large fluctuations in velocity make little difference. Over periods longer than a few weeks, wind velocities are monotonously predictable. It is for intermediate periods that an unexpected calm can be disruptive. But all power plants have periodic outages, planned and unplanned. The technology for storing energy in batteries, fuel cells, and pumped storage facilities promises to remove one of the last obstacles to the effective use of wind power (figure 7-1).

Although the seasonal nature of the wind is a disadvantage in some places, in others, it is in phase with power needs. In Denmark, for example, where many homes are heated electrically, the wind blows fastest and most steadily during the coldest months when peak demands for electricity occur.

Many countries with great need for energy have excellent potential for wind power. Conditions are favorable along the northwestern and southwestern coasts of Africa and probably along the entire western coast, as well as in Madagascar. There are potential sites along the Red Sea. In the interior of the continent things are less promising, but there are some good sites. In Asia, winds are light and variable during spring and fall, more powerful during winter and summer. Pakistan, India, Israel, and coastal Arabia have many good sites that can supply domestic power the year around with other uses perhaps confined to certain months. Australia

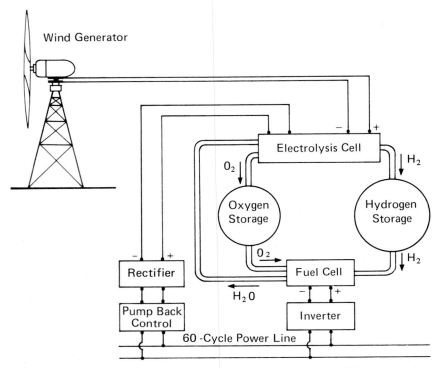

Figure 7-1.   To harness wind power effectively one must be able to store the energy captured when the wind blows and release it more or less continuously. One scheme would be to use the electricity generated by the wind to decompose water electrolytically. The stored hydrogen and oxygen could than be fed at a constant rate into a fuel cell which would produce direct current. This current would be converted into alternating current and fed into a power line. Off-peak power generated elsewhere could also be used to run the electrolysis cell when the wind was deficient. (Redrawn by permission from C. N. Summers, "The Conversion of Energy," *Scientific American* 224, no. 3 [1971]: 158.)

has good potential along most of its coast. North America could develop wind power along the Atlantic and Pacific coasts and in the plains states east of the Rocky Mountains. In South America there are excellent sites along much of the west coast and in southern Argentina.

Because windmills require no fuel and are relatively maintenance-free, their operating costs are low. Their useful life varies — it may be as short as fifteen years, but frequently is much longer. There is no pollution and no damage to the environment. A minimum of land is required for small installations. Most people find that a windmill has functional beauty. Certainly no other power-producing device except the old-fashioned waterwheel is a favorite theme of artists and photographers.

Initial cost is a problem if a great many units are required. Windmills cannot compete economically with a 1 million kilowatt nuclear power plant. On the other hand, very few rural communities could use a 1 million kilowatt nuclear power plant, so the question of scale has two faces. A 1000 kilowatt nuclear power plant could not compete economically with a windmill.

There are advantages to small, local facilities for generating electricity in spite of the trend in the United States toward gargantuan centralized installations. Wasteful, expensive, and environmentally destructive long-distance transmission of electricity becomes unnecessary, and the impact of a power failure is on the same small scale as the power plant itself.

For the industrialized world, wind power can ease the joint problems of fuel shortages and pollution. In a prototype solar-wind house in Arizona, heat is provided by the sun and there are plans for air conditioning. The house is electrified by the wind, which stores energy in batteries. Life is not primitive in this house. Luxuries like food blenders are easily accommodated.

For great areas of the undeveloped world, the wind can supply an increment of energy that will greatly enhance the elusive "quality of life." Industrialization was painful for the Western world. It threatens to be

Left, a beautiful old sail windmill in use on Myconos Island off Greece. (Photo by Margaret H. Irwin.) *Right,* a Jopp windmill in use in Minnesota. This is a 3 kilowatt, 110 volt d. c., direct drive unit. Many units similar to it were built and used during the 1930s, 1940s, and 1950s. (Photo by Don Marier.)

even more painful for many ancient cultures if old ways are suddenly destroyed without adequate preparation for new. Perhaps countries only now beginning to industrialize can prevent some of the misery suffered by the West by adapting native handicrafts to local small-scale manufacture of modern products. In many cases wind power can be adequate for this.

Another source of inanimate energy which has been known for centuries is the tides. Derived from the energy of the earth-moon-sun system, tidal energy has a future as long as that of the system itself. Tidal energy is expended at a rate of about 3 billion kilowatts. Most of this expenditure is in the oceans, especially in shallow seas, bays, and the mouths of rivers where fresh and salt water mix. In these areas the difference in sea level between high and low tide ranges from less than 3 feet to more than 30. To harness the energy of ocean tides requires only the damming of a partially enclosed basin and the installation of facilities for generating electricity.

Tidal electric power is a special kind of hydroelectric power, depending on the alternate filling and emptying of a dammed basin rather than on the one-directional flow of water. The amount of energy available at a site depends on the range of the tide and the area of the enclosed basin, and can be calculated if these values are known. The maximum rate of energy flow at the most promising sites in North America, South America, and Europe is 64 million kilowatts.[4] If we assume that 20 percent of this (a liberal allowance) can be converted to electricity, the potential for tidal electricity is about 13 million kilowatts — less than 2 percent of the world's installed electric capacity. But if you happen to live near one of the tidal sites, the tides could give you all of your electricity without poison and forever.

Exploitation of tidal energy has barely begun, although small tidal mills date back to the eleventh century. An eleventh century mill at Dover, England, reportedly caused disaster to ships because of its disturbance of the sea. The world's first and only major tidal electric plant began operating in France in 1966 at the mouth of the Rance River, where the tidal range averages 28 feet. Proposed modifications will increase the original 240 megawatt capacity to 320 megawatts, using a creditable 24 percent of the available energy. A smaller, experimental plant has been designed in Russia.

The greatest tidal ranges in the world, exceeding 50 feet, occur in the Bay of Fundy off the eastern coast of Canada. The Canadian government, after investigating the feasibility of developing tidal power in the bay, concluded that although the potential energy production would be great (more than 13 billion kilowatt-hours annually, at three sites), development is not economically feasible now. The United States reached a similar conclusion on the proposed project in Passamaquoddy Bay along the border between Maine and New Brunswick.

Conservation groups oppose tidal projects on the grounds that they will destroy natural scenery and modify coastal environments. These charges are true. But when the environmental costs of a tidal plant are compared with the costs of the alternatives, a tidal plant might prove to be the best choice.

The kinetic energy of falling water is derived from solar energy. The kinetic energy of the earth's water is the energy of all the drops of rain that fall and flow to the sea. Like the energy of winds and tides, it is constantly released whether we use it or not, and its lifetime is that of the earth as we know it.

More than two thousand years ago, man began to build waterwheels whose energy was coupled with many kinds of machines. Until the development of the steam engine, water power furnished most of the energy for industry. But hydropower had to be used where it was produced, while the noisy, dirty steam engine had the advantage of mobility. When large-scale generation and transmission of electricity became possible at the beginning of the twentieth century, the old waterwheel gave birth to the hydroelectric plant. Since then, the increase in hydroelectric power has followed a typical growth curve, about two-thirds of the present capacity having been added since World War II.

To produce hydroelectric power, a dam must be built on a river with turbines and generators at its base. The water behind the dam stores energy that would otherwise have been dissipated as the water descended over the river bed. When tunnels through the dam are opened, water pours through with great force, spinning turbines which drive generators to produce electricity.

In many ways, hydropower is the best source of electricity we now have. It requires no mining, processing, transportation, or burning of fuel. The only waste produced is frictional heat from the turbines and generators, which would have been lost anyway if the water had fallen freely. Stresses on the structural materials of the plant are minimal because no phase of power production involves high temperatures, radioactivity, or corrosive chemicals. In addition, hydroelectric plants are usually part of larger developments involving navigation, recreation, fish and wildlife management, water quality control, water supply, flood control, or irrigation. These features combine to make hydroelectricity very cheap, unless it must be transmitted over long distances. Another advantage is that a hydroelectric plant can be started and brought to full capacity quickly. Thus hydropower is being used increasingly to supply electricity at times of peak demand.

When a lake is created where there was once a river, things cannot be the same as they were before. But even though a hydroelectric plant inevitably changes the environment, it does not poison it. The change may

The TVA's Wheeler Dam on the Tennessee River. Not many such large-scale dam sites remain to be developed. The proportion of energy supplied by hydroelectric plants is likely to diminish in the future. (Photo courtesy of Tennessee Valley Authority.)

actually please many people, for it provides fishing and recreational activities where once a remote river was enjoyed by only a few.

There are problems — some with partial solutions, some with none. The reservoirs occupy large areas of land which is often prime agricultural land, and its exchange for an increment of electricity may be a bad bargain. It is also difficult to evaluate the exchange of unspoiled scenery for a recreational development. Only a few people derive satisfaction from knowing the wilderness is there, whether or not they ever see it. But when a wild river is developed it is gone forever, and we already have a lot of lakes. Technology cannot choose between producing food and producing

A new turbine being installed on one of the eleven generating units at Wheeler Dam. At 356,000 kilowatts, this dam ranks as a large hydroelectric installation, yet its output is small compared to large new nuclear plants. (Photo courtesy of Tennessee Valley Authority.)

electricity, or between fly casting for brook trout and spinning for bass. These questions must be decided by society.

Although there is no thermal pollution in the usual sense, the temperature of the water downstream from the dam can be affected. Water in a shallow reservoir becomes warmer than the river and increases the temperature downstream when it is released. Water at the bottom of a deep reservoir may be very cold. This water lowers the temperature downstream. The cold water may also be depleted in oxygen through the metabolism of microorganisms and failure to mix with surface waters. The problem of cold, oxygen-poor water can be corrected by aerating the water and releasing it from several levels to insure the proper temperature. The problem of warm water remains intractable.

When hydroelectric plants are operated intermittently, the flow of the river is grossly disturbed. Many indigenous species cannot survive the fluctuating flow. A second reservoir downstream can regulate the rate of flow, but this requires a lot of extra space and destroys an additional section of the river.

Dams may destroy the spawning grounds of migratory fish or prevent the fish from reaching them. The effectiveness of facilities for permitting safe passage of fish is limited. The cumulative impact of a series of dams on the Columbia River has greatly reduced game fish runs.

Leakage from the reservoir may cause the water table to rise, bringing with it dissolved salts and impairing the fertility of the soil. If the reservoir is shallow, a high rate of evaporation may lead to increased salinity. Sometimes these undesirable side effects can be prevented or at least minimized if they are foreseen.

At the end of 1970, 53,000 megawatts of hydroelectricity had been developed in the United States and an estimated maximum of 94,000 megawatts remained, not counting Alaska's potential of about 30,000 megawatts. But these statistics are misleading. In reality, hydroelectric power may have gone about as far as it can go in the United States. Development of a hydroelectric site is expensive and is seldom undertaken except in conjunction with other programs. To many people, as they contemplate the loss of America's few remaining wild river gorges, the best program is to leave them alone. The best sites are already in use. Most of the remaining ones are in the Pacific Northwest and Alaska, but even here good sites are limited. Economic and other factors will preclude developing many of them. Legislation may prevent use of others (for example, the Wild and Scenic Rivers Act, Public Law 90-542). Plans for Alaska are viewed with particular concern because of the vulnerability of the arctic and subarctic environment and because development threatens some of North America's few remaining major wildlife habitats.

Hydroelectric plants supplied about 15 percent of the electricity used in the United States in the early 1970s. Although the amount of hydroelectricity will continue to grow slowly for several decades, due partly to increases in capacity of existing plants, the percentage it contributes will probably decrease to around 7 percent by the end of the century. Importing hydroelectricity from Canada, which has an enormous undeveloped potential, cannot be a long-term arrangement because in ten years Canada's own growth will require what is available as surplus today.

The situation for the world is more promising. About one-tenth of the world's potential hydropower has been developed. Even more significantly, the global capacity for hydropower is almost four times as great as the total installed electric power capacity.[5] The greatest potential is in Africa and South America, the two continents which are deficient in coal.

Hydropower cannot be a long-term solution to the world's energy problems, although it can contribute substantially to the development of nations that have the greatest need for energy. The problem is silt. All rivers carry silt, tiny soil particles that are swept along with flowing water but which settle out in calm water. Thus a river's burden of silt will be dumped into the reservoir at the hydroelectric plant, a process called siltation, and the reservoir will gradually fill up. When the storage capacity of the reservoir is destroyed, the energy of the waterfall at the dam site will remain — smaller in amount, and subject to fluctuations in the river's flow. A small reservoir, fed by a river heavily laden with particulate matter, may fill up in fifty years. Other reservoirs may have a useful life of several centuries. The average is probably about a century. The dams must be maintained long after their usefulness has passed, for the accumulated silt cannot be permitted to wash suddenly downstream.

The fossil fuels, winds, water power, the food we eat, and the tides are gifts to the earth from the sun and the moon. But the earth has its own energy, for mountain building and earthquakes, thermal processes and mechanical deformations. These processes have gone on at an apparently undiminished rate throughout geological time, driven by a reservoir of energy within the earth itself.

During the last century, men discovered that heat flows continuously from the interior of the earth to the surface, from which it is radiated into space. Volcanoes and hot springs gave further evidence for the earth's heat. If the earth is losing heat, the reasoning went, then the earth is cooling off — a theory compatible with the fiery birth of the planet from its parent sun. But that theory of the earth's origin is no longer in favor and there is little evidence that the earth is growing cooler despite its loss of heat. The heat generated by decay of radioactive elements in the crust and mantle is now considered sufficient to account for most or all of the heat flow and most of the geothermal processes that occur. The earth also has gravitational and rotational energy; there are phase changes from liquid to solid and changes in crystal structure brought about by the increase of temperature and pressure with depth, all of which involve changes in energy; and there are little-understood electromagnetic phenomena in the core of the earth. But the energy which directly affects man and the face of the earth is the energy of the radioactive atom. If we use the heat of the earth, we are indirectly using nuclear energy.

The average heat flow from the surface of the earth is about 8 millionths of a calorie per square inch, or about 260 calories per year. This energy would melt a sheet of ice two-tenths of an inch thick in a year's time, a job that could be accomplished by the sun in a few hours. Heat generated in the earth is lost so slowly because rocks are very poor conductors of heat; thus, the surface of a rock exposed to sunlight becomes hot while the underside remains cool.

The total amount of heat stored in the outer 100 kilometers of the earth at temperatures above surface temperatures is about 23,000,000,000 trillion kilowatt-hours — roughly 150 times the solar energy that strikes the outer atmosphere in one year. Most of this heat is stored in the normal geothermal gradient, but local concentrations occur. Geothermal heat is concentrated along global systems of oceanic ridges and continental rifts, and their associated zones of earthquake activity, volcanoes, and other geothermal processes. It escapes to the surface in geysers, hot springs, and volcanic eruptions. One belt of geothermal activity, the "ring of fire," stretches from South America through Central America and across western North America to Alaska. To the west and south it passes through the Aleutians, Japan, Taiwan, the Philippines, and New Zealand. Another follows the great mountain ranges that span Asia from northern India and Tibet to the Middle East and pass into Greece and Italy. These areas are characterized by vast reservoirs of underground steam and hot water which derive their heat from the friction of sliding rocks or from passage of groundwater through recently (during the last 10 million years or so) erupted volcanic rocks.

Hydrothermal energy, the energy of heated water, can be distinguished from geothermal energy, the energy of hot or molten rock, although both are commonly referred to as geothermal. Strictly speaking, it is hydrothermal energy that can now be exploited economically. Because the two technologies will probably merge in the future, we will refer to the earth's heat as geothermal energy whether it is manifested in hot rocks or hot water.

Hot springs have been used for centuries, for their real and imagined health benefits and as a source of various chemicals. The first production of geothermal electricity was in 1904, from a dry steam field in Larderello, Italy. The first electric power plant to exploit a hot water reservoir was built in Wairakei, New Zealand, in 1946. Other power plants using hot water for energy are in Japan, Mexico, and the U.S.S.R. The extensive Geysers project in California, begun in 1955, uses steam. In Iceland, geothermal energy has been used since 1925, mostly for space heating and water heating. Many other countries have geothermal research programs in progress.

Exploration for buried fields of steam or hot water and their development involve techniques similar to those used in the production of petroleum. It is no accident that oil companies have pioneered in the development of geothermal energy. Under ideal conditions, a reservoir of steam need only be drilled into so the steam can be brought to the surface, passed through insulated pipes to a power plant, and run through a low pressure steam turbine. Using hot water is a little more complicated. The pressure on the water can be reduced so that part of it becomes steam. Steam and water are then separated in a centrifuge-like device and the steam drives a turbine. Or the hot water may be used to evaporate a second liquid with a low boiling point, like butane or freon, which then drives the turbine. No fuel is needed. No oxides of sulfur or nitrogen and no radioactive wastes are produced. Geothermal electricity can be very cheap.

But it is not quite that simple. Geothermal steam is at a lower temperature and pressure than the steam ordinarily used to generate electricity, making geothermal plants inherently inefficient. A geothermal plant releases from two to three times as much waste heat as a plant burning fossil fuel, and about 75 percent more than a nuclear plant of equivalent capacity. If ten 1000 megawatt geothermal plants were in-

The Geysers power plant in Sonoma County, California. This plant now produces 502,000 kilowatts from eleven turbine-generator units, with four more units scheduled for completion by 1977. (Photo courtesy of Pacific Gas and Electric Company.)

stalled in California's Imperial Valley (not an impossibility), the waste heat added to the area of 1000 square miles would equal 5 percent of the total summer heat from the sun. Although this could have an effect on the local weather, severe environmental consequences are anticipated only in the immediate vicinity of the power plants.[6] There are potential uses for the waste heat, or even for the entire output of geothermal energy if it is not used for electricity. These include space heating, water heating, process heat for industry, and desalination of seawater.

Although the usual air pollutants are not a problem, noxious gases are often a by-product of geothermal wells. These include hydrogen sulfide, methane, and ammonia. The amount of sulfur released at The Geysers is roughly the same as the amount that would be released by an equivalent plant burning low sulfur oil. Elsewhere in California the amount of hydrogen sulfide released from plants on hot water fields could exceed the production of sulfur in plants burning high sulfur fuels.

Surplus water from a geothermal steam plant is likely to contain chemicals like boron, arsenic, and ammonia, making it imperative to protect local streams and groundwater from contamination. Again, problems related to hot water fields are likely to be much worse. The hot water may be concentrated brine, with almost ten times the concentration of salts found in seawater.

One means of disposing of the brine solves two problems at once. Wherever large quantities of fluid are removed from the ground, it is possible that the land surface will sink, with potentially disastrous consequences. For disposal and prevention of subsidence, waste brine is pumped back into the ground through wells which must be carefully located and constructed to prevent contamination of water supplies. In the Imperial Valley, where effects of subsidence would be serious, it will be necessary to import seawater for reinjection to supplement the waste brine. Plans for some areas consider desalting the waste water for agricultural use and recovering useful chemicals from the residue. This would not be feasible if subsidence were a problem. Waste water can be pumped back into the ground or spread on the surface, but not both.

Some people fear that withdrawal and reinjection of fluids will trigger earthquakes. It is impossible to say what the seismic effects will be, but many geophysicists believe that the probability of a major earthquake would be diminished. Withdrawal of hot fluids would warm the fault zones, allowing stresses to be relieved gradually by a creeping motion of the rocks rather than suddenly in a "stick-slip" earthquake.

As in any drilling operation that involves fluids under pressure, there is always the danger of a blowout. One of the early wells in Mexico blew out, spouting steam and salt water for days before it was controlled. A blowout similar to the oil well blowout in the Santa Barbara Channel occurred at The Geysers in 1957. Attempts to cap the well caused steam to flow through fissures in the ground. The situation was still not under control in 1972. A large-scale release of salt water into the environment could be disastrous, especially in an agricultural area. Regulation and supervision of geothermal developments must be at least as rigorous as they are for petroleum.

Finally, since the machinery for producing geothermal steam is just as ugly as the machinery for producing petroleum, geothermal developments must be counted an aesthetic minus. They may even destroy the famous geysers of the world. We do not know what effect tampering with a field of steam or hot water may have on nearby geysers, but we do know that geysers have captured the awe and imagination of man through the ages. Their sacrifice to man's energy addiction cannot be made lightly.

The world's total installed geothermal electric capacity is around 1000 megawatts, the output of a single large nuclear or fossil power plant. Past predictions that the potential for geothermal energy was negligible have been at least partly self-fulfilling. But recent studies indicate that the potential for the western United States alone may be from 100,000 to 10 million megawatts. Additional potential exists in Alaska, Hawaii, the Gulf Coast, the Appalachians, and the Ozarks. Possibilities in Russia may exceed the most optimistic estimates for the United States. Like water power, geothermal energy must be used where it is found. Fortunately, there is promising geothermal potential in a number of energy-poor countries (for example, Japan) which now depend heavily on imported fossil fuels.

It is usually said that geothermal energy is a depletable resource, and rapidly depletable at that, the lifetime of any development being perhaps fifty years. The experience of Italy suggests, however, that fifty years is a minimum figure — that with careful management the lifetime of a geothermal field may be centuries. One focus of current research is to determine the lifetime of these fields and their rate of natural replenishment. At The Geysers, productivity of an individual well begins at 7 to 15 million kilowatts and declines exponentially to between 1 and 3 million kilowatts, so that new wells must be added regularly. We do not know the life expectancy at the lower level of productivity.

If we learn to use the heat stored in rocks, the supply cannot be exhausted in the lifetime of our species. Geothermal enthusiasts envision drilling 20 or 30 miles through the earth's crust into the mantle and bringing heat to the surface, making geothermal energy available anywhere. Long before we accomplish that, they say, we will circulate water through hot rocks and will no longer have to depend on the specialized conditions that produce natural reservoirs of steam and hot water. However, attempts to do this have met with results so unpromising that the efforts were abandoned. One reason for the failure is the poor thermal conductivity of rock.

The difficulty might be circumvented through creation of "hot cavities" by underground nuclear explosions, making it possible to circulate water through the fractured rock to extract the heat. One disadvantage of this method is that the water would become radioactive. The geothermal industry is young, however, and we can only guess at the benefits and penalties its future holds.

The energy sources discussed so far in this chapter are either limited in amount or restricted to particular geographical areas. The sun, however, gives abundant energy to all the earth. The earliest myths and legends show man's awareness of his debt to the sun. Solar energy has been used for centuries for drying food, fuel, and other materials and for obtaining salt from brine. Through trial and error and time, native architectures have evolved both to take advantage of the sun and to give refuge from it.

With technological sophistication have come sophisticated schemes for using the sun. The goals are tantalizing because solar energy is widely available, immense in quantity, non-polluting, and free for the taking. But so far the harnessing of solar energy has not been one of technology's outstanding success stories. This is due partly to the dilute and intermittent nature of the energy, but partly also to deeply rooted cultural preferences and the inertia of established industrial and economic patterns.

It is interesting to compare the technological problems delaying the use of solar energy with those related to nuclear fusion. For solar energy, all necessary knowledge and techniques exist and need only to be brought together from their diverse sources. We can calculate with fair accuracy how much a particular application of solar energy would cost under present conditions, and be certain that this cost will decrease, as cost always does, when the components are mass-produced in the future. We can use solar energy now, in many ways, whether or not we choose to do it. Not so, for fusion. We cannot estimate the cost of building a fusion reactor because we have only the foggiest notion of what we are trying to build. There is no guarantee that we will be able to design a safe — or even functional — fusion reactor.

The total amount of solar energy striking earth's atmosphere in a year is about 35,000 times the energy used annually by man. Its average intensity measured on a plane perpendicular to its path is 0.12 kilowatts per square foot, at the outer limits of the atmosphere. This number is known as the solar constant. Energy is received at this maximum rate only when the sun is directly overhead, because only then does the hypothetical perpendicular plane correspond to the "surface" of the atmosphere. At other times, the rate depends on the angle at which radiation strikes, which in turn varies with latitude, time of day, and season of the year. At any time, about half the earth is receiving no direct radiation at all.

Not all the radiation striking the atmosphere continues directly to earth. Some is reflected back into space or absorbed; some is reflected back and forth a number of times before reaching the earth as diffuse radiation rather than beam radiation coming on a direct line from the sun. As a result, the maximum intensity of solar radiation at the earth's surface is about 0.11 kilowatts per square foot, and this is encountered only on clear days at low latitudes around noontime. The total energy received under these ideal conditions is from 0.54 to 0.72 kilowatt-hours per square foot per day.

In addition to predictable variations in insolation related to the days and seasons, unpredictable variations occur due to weather and atmospheric conditions. On a clear day up to 90 percent of the incoming radiation is of the beam type, while on an overcast day it may be 100 percent diffuse. Air pollution is a significant factor. It has reduced the amount of sunlight reaching Washington, D.C., by 20 percent during the last fifty years. Plans to collect solar energy in the southwest may be threatened by air pollution.

When all losses are accounted for, the amount of solar energy intercepted by the United States is 600 times the amount of energy it uses. If we could use this solar energy with an overall efficiency of 25 percent (optimistic, but not impossible), it would provide 150 times the energy we used in 1970.

The two drawbacks to the use of solar energy are that it is dilute and must therefore be collected over large areas, and it is not only intermittent but capricious and must therefore be stored so that it is available when needed. Solutions to these problems depend on the uses for which the energy is intended. In general, the seriousness of the collection problem increases with the scale of the project.

Perhaps the oldest use of solar energy is for drying foods. The material is simply spread out in the sunshine without regard for dirt, insects, and decay. Simple improvements on this ancient practice protect the food from most of the dirt, insects, and decay. More elaborate schemes incorporate a solar collector which supplies hot air to a drying unit. It is possible to construct simple, inexpensive, and easy-to-use solar driers that yield a high quality product and, if widely used, would permit improved utilization of food in developing nations.

Akin to solar drying is solar evaporation for the production of salts, another ancient practice which is used widely today where evaporation exceeds rainfall. Evaporation ponds are used in many industrial processes. They are simple and effective, although they can do much harm to migratory waterfowl that succumb to the temptation to stop for a while and rest.

The most widespread and successful modern use of solar energy is for domestic water heating. Areas between 45° N latitude and 45° S latitude having more than two thousand hours of sunshine per year are potential candidates for solar water heating. This solar belt houses most of the world's people. Canada, northern Europe, and northern parts of the U.S.S.R. are the main areas lying outside of it.

A solar water heater for family use consists of a collecting surface with an area ranging from 10 to 30 square feet, an insulated storage tank, and pipes for circulating water. It requires minimal maintenance and supplies hot water at temperatures up to 200° F. At the end of 1960, about 350,000 solar water heaters were in use in Japan; this number had more than doubled by 1970. Their popularity had already reached a peak in 1960 and was declining in the United States, although some 25,000 were still operating in Florida. Recently, the United States has seen a revived interest in solar water heating, especially for swimming pools. Australia has a thriving solar water heater industry. A large number of units are installed in Israel. Other countries have been slow to adopt solar water heating either because there has been little demand for domestic hot water or because other forms of energy have been abundant. In the United States, the natural gas and electric power industries would probably have opposed development of the competitive industry.

Solar distillation is similar to solar evaporation, except that the product is water instead of salt. The two processes can often be combined. Solar radiation is admitted to a covered brine basin. As water evaporates, the vapor condenses on the covers and flows into collection troughs leading to a storage tank. This method was first used in Chile in 1872, to provide drinking water for animals working in the nitrate mines. Today, solar distillation supplies water for isolated areas in Australia and small communities along the Mediterranean and the Caribbean. The process is almost ready for extensive commercial application. Large distillation projects, however, encounter the same problems of economy and scale that plague all large-scale uses of solar energy. For the present, solar distillation is an attractive method for providing fresh water for families and small groups, but it could not supply the large amount of water needed, for example, in irrigation.

Use of solar energy for cooling and refrigeration is especially appealing because the most energy is available when and where the need is greatest. Refrigeration is vitally important in poor countries, where food spoilage compounds the problem of food shortage. Although this technology is in its infancy, only lack of effort holds it back. Introduction of refrigeration into developing countries may meet resistance because the complexity of the machines makes them useless to their intended users and because there is a well-founded reluctance to eat food that is more than a day old. Nevertheless, the benefits from solar refrigeration would be immense.

Solar air conditioning is a possibility for both hot and temperate climates, but its development, too, is in the early stages. Developing countries would benefit from air conditioning of offices, factories, and hospitals, but it would be a mistake to discard traditional architecture in favor of ill-adapted modern buildings that need air conditioning. In temperate climates, systems for solar air conditioning could be combined with systems for space heating.

Solar air conditioning systems work either on the principles of radiative cooling or of refrigeration. If coupled with solar heating (either of space or water), radiative cooling is simple, effective, and economical where the atmosphere is dry and clear. For heating, rooftop collectors absorb radiation during the daylight hours, transferring it to a heat storage system from which heat is available as required. At night, the collector can act as a radiator, radiating heat which has been removed from the interior of the building during the day and stored in a second heat storage system. Where meteorologic conditions are not favorable for radiative cooling, solar energy can be used to operate a heat pump, transferring heat against a temperature gradient. A solar air conditioner of this type works very much like a gas refrigerator. The technology is rather expensive, but future improvements and needs can be expected to render solar refrigeration more practically and economically attractive.

Several types of solar ovens and cookers have been on the market for decades. This use of solar energy seems simple and rewarding, but it is an example of a good idea that ran afoul of the idiosyncrasies of its supposed beneficiaries. Despite extensive field trials in India, Mexico, and Morocco, the cookers have not gained social acceptance. Better ways to store energy, to permit indoor or evening cooking, might solve this problem.

Between 20 and 30 percent of the energy used in the world is for space heating. If solar energy could contribute a substantial fraction of this, it would represent a huge saving in fuel. Space heating with solar energy is feasible in climates as cold as that of Boston. Figure 7-2 shows one arrangement for solar heating and cooling. So far, however, neither government nor industry has promoted the idea. The coal and oil industries have a corner on most innovations in energy technology, from production of synthetic fuels to exploitation of geothermal energy, but solar energy is outside their area of expertise and they are not interested in it. Building contractors are concerned with mass-produced housing at the lowest cost. They cannot be expected to experiment with expensive new systems that require fundamental changes in materials and design. The public has not been educated to appreciate the ultimate savings in a home with a high initial cost but low maintenance costs, such as a "solar house" would be.

A number of solar houses have been built and operated with varying degrees of success. Collectors on the roof transfer heat into a stream of water or air. The heat is carried to a water tank or gravel bin for storage,

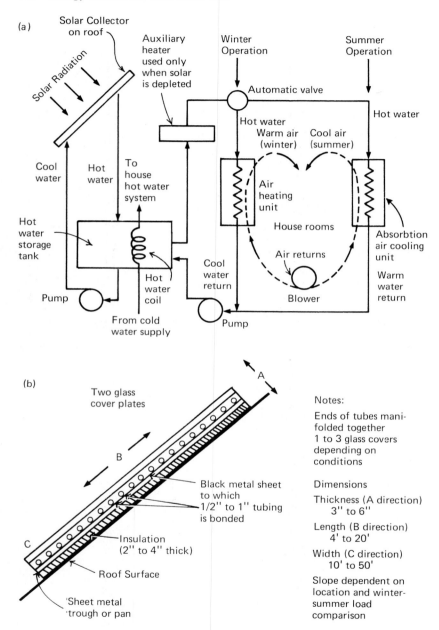

Figure 7-2. a, Solar home heating, cooling, and hot water. Schematic diagram of one alternative. b, Solar collector for home heating and cooling system. Diagram of one alternative. (Adapted from NSF/NASA Solar Energy Panel, *Solar Energy as a National Resource.* Report No. NSF/RA/N-73-001, 1972.)

Three houses with solar heating. Note the varied architectural treatment of the solar collectors. *Top,* a solar-heated house near Odiello in southern France. *Center,* the Lof house in Colorado. This house has been operating successfully since 1959. Solar panels may be seen on roof. *Bottom,* MIT solar house IV. This experimental house was built as part of the Solar Energy Research Program at MIT. (Photos courtesy of Solar Energy Laboratory, University of Wisconsin.)

from which it is circulated to the living area as needed. The houses are constructed with careful attention to reflective or absorptive properties of building materials, location of windows, insulation, and so on, to minimize the amount of heating required. This is sound building practice no matter what the source of heat, but it is often neglected. Economically, solar energy is already less expensive than electricity for space heating in most parts of the country. Rising prices for gas and oil would make solar heating the cheapest method almost everywhere.

A solar heating system would be grossly overdesigned for normal operation if it could supply heat for an extended sunless period during the coldest months. Therefore, solar houses have supplementary heating systems which provide from 20 to 70 percent of the heat. This gives a saving in fuel or electricity ranging from 30 to 80 percent. The most satisfactory design for the Boston area supplies about half the heat from the solar system.

Solar energy is ideal for small, decentralized applications. But industrialized society also needs large amounts of concentrated energy. Solar energy is one of the only three candidates that can fill this need in the long term.

On a large scale, solar energy can be collected either on the surface of the earth or on satellites orbiting the earth. The use of satellites may be an extraordinarily complicated and expensive way to circumvent the inconvenience of cloudy weather and the cycle of days and nights. According to one proposed scheme, a solar power station with a panel of solar cells would be constructed in space from a space shuttle. The orbiting station would receive sunlight continuously and convert it to electricity with an efficiency of about 15 percent. This electricity would be converted to microwaves (electromagnetic radiation at frequencies between conventional radiowaves and the far infrared) with an efficiency of 85 percent. In this form it would be transmitted to the earth, penetrating the atmosphere, clear or cloudy, with essentially no loss. An earth-based receiving station would reconvert the microwaves to electricity with an efficiency of perhaps 70 percent. To collect enough energy to supply the power needs of New York City, the orbiting panel of solar cells would have to be 25 square miles in area. The receiver on earth would occupy 36 square miles, in order to keep the radiation at a biologically safe level. If the projected electric power production for the year 2000 were to be supplied by space stations of this type, we would need 250 of them.[7]

Most proposals for solar electricity involve covering large areas of desert with collectors of solar energy. These collectors are of two main types. Flat plate collectors are basically blackened surfaces which absorb radiation, both beam and diffuse, and convert it to heat. The heat is transferred to a stream of air, water, or other fluid and carried to where it will be used or stored. Except for a few experimental solar furnaces, all of today's practical solar energy systems use this type of collector.

The technology of flat plate collectors is advanced, but it is uncertain how much mass production will be able to reduce their cost. The maximum obtainable temperatures are rather low, leading to inefficiency and a high level of thermal pollution. The maximum yield of electricity from a power plant using flat plate collectors would be about 30 megawatts per square kilometer of collector surface.[8] Since the collectors must be spaced in such a way that they do not shade each other, the total area occupied would be between 30 and 40 square miles for a 1000 megawatt plant. Seasonal fluctuations in energy can be minimized and overall efficiency of the collectors increased by tilting the collectors toward the equator at an angle equal to the latitude.

The alternative to simple flat absorbing surfaces is a system of lenses or mirrors which focus solar energy on a receiver smaller than the collector, permitting very high temperatures to be realized and reducing the area from which energy is lost by radiation. Ratios of intensity of radiation on the receiver to intensity on the collector can range from 2 or 3 up to more than 10,000. To obtain any but the lowest of these, the collectors must track the sun across the sky. The special mountings and controls for tracking add to the cost and complexity of the system, but without them the use of focusing collectors is not justified.

Only the beam component of radiation can be focused. The inability to use diffuse radiation accentuates the differences in energy available in clear as compared to cloudy weather and in different seasons of the year. In Texas the loss varies from about 10 percent in summer to as much as 42 percent in winter.[9]

Scientists have proposed a variety of designs for solar power plants using focusing systems. One of these would focus energy on a boiler with square aluminum reflectors, 10 to 16 feet on a side.[10] The reflectors would be individually mounted with automatic controls for tracking the sun. If the total projected energy requirements of the United States for the year 2000 were to be met by this — or any similar — system, between 1 and 2 percent of the country would be covered by collecting devices. (By comparison, an area larger than this has already been turned over to the automobile, in the form of parking lots and roads; a tally of the land occupied by fossil fuel and nuclear plants must include land ravaged by the mines that feed them; and half a million square miles are devoted to the inefficient energy industry called agriculture.) About 5.75 square miles of collector surface could provide energy for a 1000 megawatt power plant, assuming 30 percent efficiency of conversion. Allowing for spacing of the collectors, about 16 square miles of land would be involved, or a square 4 miles on a side.

Another proposal takes advantage of mass-produced plastic Fresnel lenses. These lenses have the optical properties of spherical lenses, but they are telescoped in a way resembling a collapsible drinking cup, at great savings in size, weight, and materials.[11] A third system favors exploiting recent and, hopefully, future advances in the technology of selective surfaces. The selective surface might be a material that absorbs radiation at wavelengths shorter than about 1 micron in the infrared but is transparent to longer wavelengths. This would allow absorption of most of the solar spectrum while minimizing losses from radiation. Selective absorption could also be based on optical interference between layers of material, regardless of the intrinsic optical properties of the materials themselves.[12]

After solar energy is collected, it must be converted to a useful form, and here the problem of storage for sunless periods arises. One solution, suggested with increasing frequency, is to use the energy for producing hydrogen. In many ways, hydrogen is an ideal fuel. It can be produced from water, and on burning it yields only water as a combustion product. With minor modification of existing storage facilities, pipelines, and burners, it could be stored, distributed, and used as natural gas is now. Since large solar plants will be located in desert areas far from population centers where energy is needed, a fuel that can be moved efficiently and relatively cheaply by pipeline has an advantage over electricity which is transmitted inefficiently and at great cost.

Hydrogen could be produced by thermal dissociation of water or by electrolysis. At high temperatures, water dissociates into hydrogen and oxygen. If one of these products is continuously removed so that water cannot be reformed, the reaction will continue rather than coming to equilibrium. If a practical way can be found to remove the small percentage of hydrogen formed (two atoms, or one molecule, for every 999 molecules of water vapor at 2700° F), we could expect a yield of about half an ounce of hydrogen per day from each square foot of collector area. Oxygen would be a valuable byproduct.[13]

If solar energy were first converted to electricity through a conventional steam cycle or some advanced process, the electricity could be used to decompose water by electrolysis. Here the products, hydrogen and oxygen, are easily recovered, because they form at opposite electrical poles. Although the electrolytic production of hydrogen from solar energy and water is an attractive prospect, the amount of electricity required is very large. One estimate is that if we were to produce an amount of hydrogen equivalent in energy value only to the natural gas consumed in the United States, we would have to quadruple our electrical generating capacity.[14]

Like other forms of energy, solar energy is not entirely free. The process of converting it to useful form has a price, part of which is the large amount of space and materials required for conversion. But when we look at the seashores, hillsides, waterways, and agricultural land that are being sacrificed to the production of energy, devoting a piece of the desert to this purpose does not seem an unreasonable alternative.

The desert is fragile in the same sense that the arctic environment is fragile. Its ecology could easily be upset. The greatest threat to the desert environment is probably not a few solar power plants in themselves, but the influx of people that would accompany the production and availability of power.

Other possible consequences of using solar energy relate to upsetting the heat balance of the earth. Thermal pollution of local water resources could be severe if water-cooled steam electric generating systems were used. Trapping large amounts of solar energy in the desert and transporting it elsewhere for use would alter the distribution of solar energy on earth. By blackening the desert with solar collectors or collecting radiation in space, we are causing the earth to absorb some energy that otherwise would have been reflected. But the desert itself will not heat up because we are taking the energy away; and the areas that use it are already heating themselves with the inevitable waste heat from other sources of energy. If there is a threat to global climate and heat balance, it will not be from the use of solar energy, but from the intemperate use of energy in any and all forms.

In conclusion, we may well take a lesson from ecology. The United States is planning to construct great clusters of nuclear power plants, each unit of which will be large by today's standards. They will supply power to areas covering many states, as we grow toward an all-electric economy. Perhaps, before committing ourselves to this course, we should remember the lesson from ecology that teaches us that diversity is the key to survival. A diversity of species lends stability to the environment, genetic diversity enhances the adaptability of a species, and physiological diversity increases the chances of an individual for survival. Yet we behave as if the human species and its cultures were exempt from this rule. We have already experienced what a power failure can do to New York, but we are asking for these crises on an even larger scale. A better plan might be to exploit as wide a variety of energy sources as possible, each according to its best use. Regional energy plans should suit local needs and the availability of resources.

If something goes wrong with your solar heating system, you can go to your neighbor's house to keep warm. If something goes wrong with the central power plant 500 miles away, there is nowhere to go.

References

1. J. Solomon, "The Social Redemption of Pure Garbage," *The Sciences* July-August 1972: 13-15.
2. J. Reynolds, *Windmills and Watermills* (London: Hugh Evelyn Limited, 1970).
3. United Nations, Proceedings of the United Nations Conference on New Sources of Energy, *Wind Power* vol. 7 (New York, 1964).

4. M. K. Hubbert, "Energy Resources," in *Resources and Man*, National Academy of Sciences-National Research Council (San Francisco: W. H. Freeman and Company, 1969), pp. 157-242.
5. *Ibid.*
6. M. Goldsmith, *Geothermal Resources in California, Potentials and Problems.* Environmental Quality Laboratory Report No. 5 (Pasadena: California Institute of Technology, 1971), pp. 32-33.
7. C. Summers, "The Conversion of Energy," *Scientific American* 224, No. 3 (1971): 158-159.
8. H. Hottell and J. Howard, *New Energy Technology: Some Facts and Assessments* (Cambridge, Mass.: The MIT Press, 1971), pp. 331-345.
9. *Ibid.*
10. A. Hildebrandt *et al.*, "Large-Scale Concentration and Conversion of Solar Energy," *Eos: Transactions, American Geophysical Union* 53 (1972): 684-692.
11. N. Ford and J. Kane, "Solar Power," *Science and Public Affairs* 27, No. 8 (1971): 27-31.
12. A. Meinel and M. Meinel, "Is It Time for a New Look at Solar Energy?" *Science and Public Affairs* 27, No. 8 (1971): 32-37.
13. Ford and Kane, *op. cit.*
14. D. Gregory, "The Hydrogen Economy," *Scientific American* 228, No. 1 (1973): 13-21.

8

ELECTRICITY

*Poison is in everything and no thing is without poison. The dosage
makes it either a poison or a remedy.*

Paracelsus

Electricity is not a primary source of energy, but merely a convenient and
versatile form of it. Electricity must always be derived from another source
of energy — fuels, energy of motion, electromagnetic radiation, or nuclear
energy. Since electricity is produced from one of the primary energy
sources, it creates new markets for all of them at the same time that it takes
their traditional markets away. For example, the current trend is toward
electric space heating in homes and electric furnaces for the production of
steel. Only the internal combustion engine seems to be holding its own
against electricity. Because the conversion of energy from one form to
another is never completely efficient except, in some cases, for the conversion to heat, the use of electricity is generally an inefficient use of energy.
The burning of fuels for generation of electricity which is used for heating
is a very wasteful process. Nor is the use of batteries a step toward energy
conservation. A battery is merely a device for storing energy. Several times
as much energy may be used in the manufacture and charging of a battery
as can be derived from its use.

In 1970, 23 percent of the energy used by the United States was used
for the generation of electricity. There are interesting problems and implications hidden behind this uninteresting statistic. Since World War I
the use of electricity has been doubling at eight- or ten-year intervals, with
a similar rate projected for the future, while energy use in general has a

doubling time about twice as long. If this trend continues, we will have an all-electric economy by the year 2040. All the technological, environmental, political, and economic problems of all the energy industries are involved in the electric power industry. And the generation, distribution, and use of electricity pose their own unique set of problems and promises.

The increase in generating capacity in the United States has been matched by an increase in the maximum size of generating units and by the trend toward using multiple units in a central power station. In 1920, the maximum generating capacity of a single unit was 50 megawatts. By 1930, it was 200 megawatts. In 1960, the giants had a capacity of about 400 megawatts and by 1970 some 1300 megawatt units were on order. Plans for 1990 assume that individual units of 2000 megawatt capacity will be available.

Electric utilities once had free rein in selecting sites for power plants. Today, almost every proposal for a new site meets bitter opposition from someone, and construction of new plants is delayed for many months,

Coal-fired power plants require large coal piles and land for them. This plant, the TVA's John Sevier plant, has a capacity of 823,000 kilowatts and burns about 2 million tons of coal a year. (Photo courtesy of Tennessee Valley Authority.)

pending court decisions. Meanwhile, the electric power industry insists it will have to build about 500 new plants during the next twenty years, and this tabulation does not include "small" plants with generating capacity below 500 megawatts.

A proposed site becomes a battleground, most of the attempts to come to a mutual understanding being thinly veiled attempts of each side to convince the other. On the one side are those who would block any expansion of electric power partly on ecologic, partly on emotional grounds. On the other side, the fatalistic attitude has been summarized as follows:

1. There is no current policy or sentiment for modifying the economic growth of the country.
2. There is no experience whatever in maintaining or developing growth of an industrialized economy with restricted amounts of electricity.
3. . . . there is more evidence that general limitation of electrical supply might either curtail the growth of the economy or impede some major objectives of environmental improvement.[1]

There are fallacies in both of the extreme positions, as there always are in extreme positions. Because we cannot change our habits immediately, or stop all growth, we will continue to need increasing amounts of electricity for the next few decades. However, consideration of the results of uncontrolled growth suggests that it may be suicidal not to begin immediately to do something about the situation described in the quotation above, rather than accept it as inevitable.

The United States will continue to need new power plants. We will discuss some of the requirements that must be satisfied and dilemmas that must be resolved in selecting sites for them.

Good highway access is necessary for construction and maintenance. Access by rail or water is also desirable for delivery of equipment and, in some cases, fuel. If adequate roads and railroads do not exist, their cost must be reckoned in the total cost of the plant. Possible seismic activity in the area must be considered, as must meteorologic hazards like tornadoes, hurricanes, ice storms, and sand storms. Design standards for a plant must reflect the probability of these natural phenomena.

An adequate and dependable supply of cooling water is extremely important, and is the source of much of the conflict over sites. Increasingly, the electric power industry is competing with other industry for sites, and all are meeting opposition from those who have other plans for our limited water resources.

New power plants will require large amounts of land, for the power house itself, equipment for controlling air pollution, fuel storage, rail, barge, and truck terminals, and transmission access and switchyards. Table 8-1 shows the land requirements for 3000 megawatt fossil-fueled and

Table 8-1.   Land Requirements for a 3000 Megawatt Power Plant

Type of Fuel	Number of Acres	Comments
Coal	900-1200	Includes about 40 acres for coal storage and 300-400 acres for ash disposal.
Oil	148-350	Includes onsite fuel storage.
Gas	100-200	Assumes pipeline delivery, with only modest onsite fuel storage.
Nuclear	200-400	Includes recommended "exclusion zone," in which the reactor licensee has authority to determine all activities. Surrounding the 200 to 400 acre site should be a zone of low population density whose inhabitants can be protected or evacuated in an emergency. These restrictions will probably be relaxed as experience with reactors increases.

Source: U.S., Office of Science and Technology, Executive Office of the President. *Considerations Affecting Steam Power Plant Selection.* Washington, D.C., 1968.

nuclear plants. If a cooling pond is incorporated into the cooling system, an additional area of 4 to 20 square miles would be needed.

There is a geographic pattern in the type of plant that will predominate. The Northeast and Midwest are leaning heavily on nuclear power. Almost all the new plants in the South Central Region will be fossil fueled. Major hydroelectric additions will be in the Northwest, with large amounts of pumped storage anticipated in the Northeast and Southeast. The Northwest, however, which traditionally has depended on hydropower, will soon be forced to shift toward thermal power for lack of hydro sites. This pattern reflects the need to guarantee an adequate fuel supply for the life of the plant, as well as other practical and economic realities.

The most fundamental question in siting a power plant is whether to locate it near or far from the population centers it will serve. A corollary of this, for fossil fuel plants, is the question of whether to transport the fuel to the power plant or build the power plant where the fuel is produced. Neither question has an easy or obvious answer.

For those who raise the point that the people who use the electricity should pay the environmental price for it, there is the counterpoint that some areas, such as Los Angeles, may not be able to afford the environmental price — and one cannot simply say, then let them do without. The argument for locating major power plants at great distances from population centers contends that there the environment is better able to assimilate the various kinds of pollution resulting from generation of electricity, and, by serving a number of load centers, one power plant can take advantage of economies of scale. Against this argument is the increased economic and environmental cost of transmission, the fact that many of the potential uses of waste heat cannot be realized, and the reluctance of many people to destroy any remaining wilderness.

Perhaps the most publicized problem created by the growth of the electric power industry is the problem of waste heat. When we say that the efficiency of a power plant is 33 percent, we are saying that for every kilowatt-hour of electricity it generates, 2 kilowatt-hours of energy are wasted in the form of heat. Because of the inescapable inefficiency of a heat engine, which is imposed by the second law of thermodynamics, the

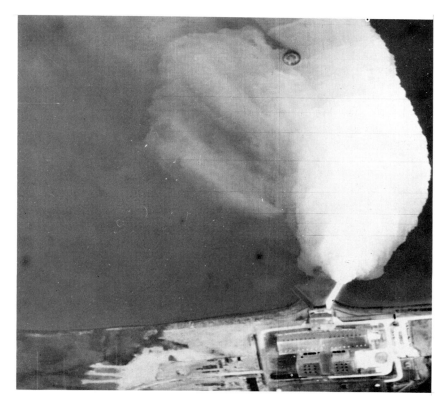

A plume of warm water discharged from a nuclear power plant on Lake Michigan. This infrared image shows the warm water as a light area in the cool lake. The thermal plume shifts in response to winds and currents. The dark circular feature well out in the plume is the water intake. (Photo courtesy of Marine Studies Center, University of Wisconsin.)

production of waste heat is inevitable and we will have to live with large amounts of it so long as we generate electricity by means of a steam cycle.

Only recently has heat come to be considered a pollutant. The electric utilities boast that they were concerned with minimizing waste heat long before the environmentalists had ever heard of it. The reason for their concern was economic. They wanted to get as much electricity as possible from each pound of fuel. The amount of heat released at a particular place was usually small compared to the total energy flow in the area and the total amount of energy used by man was negligible compared to global energy flow. Both situations are changing rapidly.

No matter where or how the waste heat is disposed of, it eventually ends up in the atmosphere, from which it is radiated into space. The possible long-range effects of discharging large amounts of heat into the atmosphere will be discussed in chapter 10. For now, we will consider only the effects of waste heat on water resources and what can be done to alleviate thermal pollution.

Power plants using steam cycles discharge waste heat when steam condenses to water in the condenser. This step is necessary because the efficiency of generating electricity depends on a sharp drop in pressure from one side of the turbine to the other. Waste heat is conventionally transferred to cooling water which, after passing through the system once (hence the name once-through cooling) is discharged to the source some 10° to 25° F warmer than when it began. The only conceivable alternative to water for cooling on this scale is air, which absorbs only about 0.025 percent as much heat as an equal volume of water. From the standpoint of engineering, there is no alternative to water as a coolant. The thermal discharge from a power plant is shown on page 151.

Each day, about 10 percent of the flowing fresh water in the United States is used by industry for cooling. About 80 percent of this, one-third of the water used for all purposes, is used in the generation of electricity. By 1980, the electric power industry will require 17 percent of the nation's flowing fresh water for cooling. A 1000 megawatt nuclear power plant operating with an efficiency of 32 percent will raise the temperature of 50 million gallons of water per hour by 15° F. This is about 50 percent more heat than a fossil fuel plant of equivalent generating capacity would discharge, partly because a nuclear plant is less efficient and partly because about a fifth of the heat from a fossil plant escapes up the stack.

It is important to distinguish between heat and temperature. We have seen, for example, that water is an ideal coolant because it can absorb more heat than most substances for each degree its temperature is raised. Within limits, a power plant can discharge its waste heat to a small amount of water, raising its temperature considerably, or to a large amount with a smaller temperature increase, but the total amount of heat released is the same. Because the rate of cooling increases with an increasing temperature of the surface water and the area required for cooling the heated water to a given temperature decreases, it may be better to discharge a minimum amount of quite warm water rather than a larger amount of cooler water. It may be better to admit that there will be some unwanted side effects of the waste heat and to keep them localized so there is hope of controlling them or limiting their spread.[2] However, factors other than the rate of cooling and the area required must be considered.

In a natural lake, about one-quarter of the energy lost to the atmosphere is lost through evaporation and three-quarters is lost by radiation. Waste heat from industrial cooling disproportionately increases loss by evaporation because the vapor pressure of water, which determines the amount of evaporation that can occur, increases with temperature. The result is that about 40 percent of the industrial heat load is dissipated through evaporation, 30 percent through radiation, and most of the rest by conduction, or the direct transfer of heat from water to air at the interface between them. The exact relationship among the types of heat loss depends on the season of the year and the water temperature.

Waste heat has a variety of effects on living things. Each type of organism has a characteristic temperature range in which it can survive and, within this range, a much smaller temperature range that favors optimal growth. The temperature range is determined not only by the direct effects of temperature on the organism but by the nature of the plant and animal communities which thrive in that range. The presence or absence of food, parasites, competitors, or predators may determine the ability of an organism to survive at a particular temperature. Thus, some types of game fish vanish from warm water not because they themselves cannot live there, but because the organisms on which they feed cannot survive.

The optimum temperature for an organism is usually close to the maximum temperature at which it can survive. Some people argue that the sun is the largest contributor to thermal pollution, both in an absolute sense and in the extent of natural temperature fluctuations, which may exceed 50° F. With this reasoning they conclude that additional inputs of waste heat are harmless. On the contrary, the upper end of the natural temperature range may be so close to the lethal temperature for many organisms that an additional increment of waste heat may be an unbearable stress. The effects of waste heat vary with the season of the year and with shorter term weather conditions. Waste heat is more likely to be beneficial in cold climates and harmful in hot ones. In addition, as any aquarium enthusiast knows, rapid fluctuations in temperature, such as occur when cooling water is intermittently discharged, can be more harmful than a constant temperature above or below the optimum.

Temperature influences the behavior of organisms, from the tiniest planktonic creatures to the great game fish. The following situations are conjectural, because we have been unable to find evidence that they have actually occurred as a result of thermal pollution from power plants. However, they could occur. There is a daily vertical migration of tiny, surface-feeding animals (zooplankters) in response to diurnal cycles of temperature and light. The effects of various strategies for disposing of waste heat on this behavior are unknown, but they could be significant. Migratory behavior of some species of fish is triggered by temperature. If animals are caused to migrate by exposure to a discharge of warmed water, several kinds of potential trouble await them at their destination. They may find conditions unfavorable for spawning. Temperatures may be too low for development of eggs and survival or growth of the young. Or, the appropriate food may not be available.

In general, plant and animal species of value to people thrive at lower temperatures than nuisance species. An outstanding example of this is the bluegreen algae. Diatoms, basic elements of many food webs, flourish at temperatures between 59° and 77° F. Green algae, also important in food webs, thrive in warmer waters at 77° to 95° F. Bluegreen algae can grow within and beyond this entire temperature range. At lower temperatures they usually cannot compete successfully with diatoms and green algae, but as temperatures increase so do the bluegreens. Bluegreen algae are not part of the food webs of most animals we wish to encourage. They flourish in the absence of predators, at the expense of green algae and diatoms and, indirectly, of the animals that prey on them. When the bluegreens die, their decomposition by microorganisms further depletes the oxygen supply that has already been diminished by increased temperature. Numerous population studies have shown that if two species are competing for the same environmental niche, one tends eventually to displace the other. This is what happens in the contest between the bluegreen algae and the greens and diatoms. Once the bluegreens have become established in a region of warm water, they have the opportunity to colonize waters relatively un-affected by thermal discharges. What begins as a rather small shift in the balance among species can have far-reaching effects. This replacement of one group of species by another, followed by deterioration of the habitat, often resembles acceleration of the natural aging, or eutrophication, of an aquatic ecosystem. The rate of change is further accelerated by addition of nitrates, phosphates, and organic carbon from agricultural runoff and sewage.

Not all effects of waste heat are undesirable. Up to a certain point, an increase in temperature can result in more rapid development of eggs and faster growth of fish of all ages. There have been plans to use waste heat in nurseries for fish and shellfish. Some waters which are too warm for trout and too cold for bass and catfish might be transformed into excellent fishing grounds by thermal discharges from a power plant. There are many reported cases of improved fishing in the vicinity of such discharges (in contrast to the lack of reports of environmental disasters). If the power plant operates intermittently, however, or is shut down for repairs or as in the case of a nuclear plant for refueling, the fish that depend on the heat it provides may suffer. We must learn many things about the effects of heat on aquatic life and we must know many things about the proposed site for a power plant before we can predict the effects of its thermal discharge. Current concern is well founded because, while the environment has so far been able to assimilate our waste heat, continued exponential growth will rapidly exceed its capacity to do so.

There exists a variety of strategies for disposal of waste heat. The easiest, cheapest, and most efficient method is the once-through cooling method, by which about 80 percent of all waste heat was discharged in 1970. Warmed cooling water may be returned to the surface, where it will stay because it is lighter than the colder water below, and where it will quickly give up its excess heat to the atmosphere. The warm water can also be returned to a deeper region where the heat will be diluted very rapidly; but because it is dilute and because cooling takes place only from the surface, the heat will be retained for a long time. Many lakes become thermally stratified during spring and summer, with layers of water at different densities and temperatures. Withdrawing water from colder layers and returning it to the surface could upset the thermal stability of these lakes, leading to unpredictable ecologic effects.

All schemes for preventing the effects of waste heat on natural bodies of water involve ways of getting the heat into the atmosphere without heating the natural body of water first. All of them require compromises in cost, efficiency, and environmental effects. Systems for treating the cooling water are discussed below and comparative energy requirements for operating them are shown in table 8-2. The best plan would be to divert as much of the waste heat as possible for useful purposes.

Table 8-2.   Power Requirements for Cooling Systems (percentage of output of 800 megawatt plant)

Cooling System	Type of Fuel Fossil	Nuclear
Once through: discharge to lake, river, ocean, or cooling pond	0.4	0.6
Wet cooling tower		
Natural draft	0.9	1.3
Mechanical draft	1.0	1.7
Dry cooling tower		
Natural draft*	0.9	1.5
Mechanical draft	3.0	4.8

*Construction of natural draft dry cooling towers is prohibitively expensive.

Based on data in M. Eisenbud and G. Gleason, eds., *Electric Power and Thermal Discharges* (New York: Gordon and Breach. 1969), p. 372.

Cooling towers discharge waste heat directly to the atmosphere. In a wet cooling tower, this is accomplished through evaporation, which raises questions of local fog, increased precipitation, and depletion of the source of cooling water. A decision to use such a tower must take all these potential problems into account as must the design of the tower itself. Where topographic and meteorologic conditions are favorable, fogging is negligible, although the plume of water vapor from the tower may be visible for miles. As for excessive water loss, it is important to remember that much of the cooling from a body of water also takes place through evaporation. The evaporative loss of water from direct discharge is about 1 percent, while that from a wet cooling tower is about 1.5 percent. Because they depend on evaporation, wet towers are less effective when the relative humidity is high.

In a wet cooling tower, water from the cooling coils of the condenser trickles through loose packing at the base of the tower, where it is met by a rising current of air. In many locations, the tower can operate on the principles of a chimney, with a natural draft of air being drawn through. These towers are enormous structures that dominate the landscape (page 157). Where natural draft towers are not feasible, fans can create a forced, or mechanical, draft. Wet cooling towers approximately double the capital cost of the cooling system over the once-through, direct discharge method. Because they are expensive and are likely to create environmental problems of their own, they should not be used unless their use can be justified.

Like wet cooling towers, dry cooling towers may be either natural or forced draft. Dry cooling towers avoid problems of fog, mist, unsightly vapor plumes, and evaporative water loss. They depend on the transfer of heat to air by convection, as the cooling water passes through a heat exchanger in a closed system and never contacts the air directly. For this reason, the effectiveness of a dry tower is independent of humidity. However, its effectiveness decreases as air temperature increases. Because dry towers decrease the efficiency of the power plant and increase the capital costs of the cooling system by up to five times or more, the use of dry cooling towers is usually not justified. In the United States, there are no dry cooling systems at major power plants although there are a few small ones; there are a number of relatively large ones operating in Europe, however.

In addition to the chemicals used in the condenser, all cooling towers must use chemicals (for example, chlorine and sulfuric acid) to prevent biological growth, deposition of salts, and corrosion in the system. Discharge of these harmful materials to natural waters must be controlled.

A cooling pond is a manmade lake in which waste heat is dissipated in the same way as in a natural lake. The advantages of cooling ponds are that they are less expensive then towers and they prevent spreading

Wet cooling towers. (Photo courtesy of Tennessee Valley Authority.)

deterioration of natural bodies of water. Their main disadvantage is the amount of land they require, although they can provide opportunities for fishing and recreation if they are properly managed.

Another means of dissipating waste heat is spray canals in which water from the condenser is sprayed from banks of spraying units into a canal that carries it back to the source. Heat is lost primarily through evaporation. Spray canals require only 5 or 10 percent of the surface area required by a pond, and they are less expensive than cooling towers.

So far we have assumed that the cooling water will be used once and returned to its source, with or without prior cooling. All the systems described can also be designed to recycle the cooling water through the cooling system, replacing water lost through evaporation and exchanging an additional fraction to prevent accumulation of salts. These systems may be the best choice for sites where water supply is a problem. A recycling system using dry cooling towers would essentially free the generation of electricity from restrictions of water supply, and would greatly simplify plant siting. Unless there are improvements in current technology, however, such a system would increase the complexity and cost of generating electricity and decrease the efficiency. At least one manufacturer of turbines in the United States is considering new designs less affected by the higher backpressures imposed on the turbine by dry cooling towers. If these and other modifications are successful, the economics of non-evaporative cooling will improve and these techniques may become the solution to cooling problems for the future.

It is clear that there is no ideal solution to the problem of waste heat. Each solution is a compromise, and bears its own economic and ecologic price tag. Sometimes the situation is clearly a trade-off between the economy and the environment, but more often, one set of problems is merely exchanged for another. The best solution may be a combination cooling system, in which once-through cooling is used at times when the extra heat burden can be assimilated without damage to the aquatic ecosystems, with auxiliary use of cooling towers, spray canals, or cooling ponds under certain conditions or in certain seasons of the year.

There are many ways in which we could use energy more efficiently than we do. One of these is to use waste heat from power plants for some of the many jobs for which moderate temperatures are adequate. Already, there are plans for combination power plant-water desalination projects. Power plants could become multipurpose energy sources for urban areas, their waste heat providing energy for space heating in winter and air conditioning in summer. Combination power plant-sewage treatment plants may be feasible, although substantial research in this area is lacking. Addition of waste heat could increase the rate of microbial reactions in sewage treatment. Raising the temperature of the sewage from the usual 70° F to 100° F would, in theory, double the capacity of the treatment plant, but this has not been verified by experiment. If treated sewage were passed through a filter of coal, more than 95 percent of the remaining organic material would be removed. This combination of coal and sludge could then be burned as fuel for the power plant, the sludge being a cleaner fuel.

Waste heat could be used to de-fog and de-ice airports. It could warm greenhouses. It could be used in fish hatcheries and shellfish farms. All of these developments could come to pass, but at present, they are in the realm of technological forecasting. If they are realized, it will only be as a result of integrated, long-range planning. They will be competing with the tendency to locate power plants far from population centers where, out of sight, they supposedly will also be out of mind, and where their side effects may or may not affect the people they serve.

In the early days of electric power systems, transmission of power from the generator to the point of use was wasteful and very expensive, and power plants were always located in the vicinity of major load centers (except for some hydroelectric plants, which were spared fuel costs and could afford higher transmission costs). Transmission is still expensive, and is consumptive of space and materials; but with; development of high voltage transmission, its cost relative to the cost of transporting fuel has declined steadily, permitting more flexibility in power plant siting. In 1910, after high voltage transmission was introduced into the United States, the maximum line voltage was 110 kilovolts (110,000 volts). By 1920 it had climbed to 200 kilovolts, it was about 350 in 1940, it was past 400 in 1960, and by 1970 an 1100 kilovolt facility was being tested. Now we talk about extra-high voltage, and, for the future, ultra-high voltage.

Electric power is the product of electric current (the rate of flow of electric charge) and electromotive force (voltage, roughly analogous to pressure). From this relationship it is evident that, as voltage increases, more energy is transmitted. Increased capacity, however, is not the only advantage of high voltages. Some energy is always lost during transmission. Most of this loss is due to resistance of the wire, which in turn, is related to the current and the cross sectional area of the wire. In the transmission of any given amount of energy, the higher the voltage is, the lower the current and consequently the lower the resistance loss will be. Alternatively, in transmitting a given amount of energy at some predetermined rate of loss, the higher the voltage is, the smaller the wire can be. Thus the three related gains from high voltage transmission are increased capacity, decreased resistance loss, and decreased size of transmission lines. At high voltages, however, other kinds of energy loss occur. One of these is a leakage of energy called a corona discharge, which occurs because the air around the transmission lines becomes ionized.

Figure 8-1 shows the elements of a power system involved in transmitting electricity. Economical voltages for transmission are much higher than those at which the systems that generate and use electricity

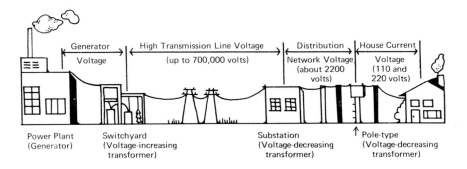

Figure 8-1. Functional diagram of the transmission and distribution system for electrical power.

Transmission stations frequently require as much land for their complex array of towers and wires as power generating plants. (Photo courtesy of Edison Electric Institute.)

can operate. In the switchyard of a power plant, a voltage transformer increases the voltage of the power from the generator to the transmission line value. The power is then distributed to various substations near load centers, where another transformer decreases the voltage before it enters the distribution network. Voltage is reduced again in the familiar transformers mounted on utility poles, from which electricity reaches the consumer at a very nearly constant voltage of 110 or 220.

Modern electric transmission towers are less ugly than the older spidery steel towers, but they still require large land areas and are frequently unpopular with landowners. (Photo courtesy of Edison Electric Institute.)

Transmission systems require a great deal of space. Lines connecting a typical 3000 megawatt plant into a transmission system at 500 kilovolts require a right-of-way of 100 to 160 acres for each mile of lines. Thus the advantages of high voltage transmission are obvious.

For aesthetic reasons, there is increasing pressure to lay transmission lines underground. At present, however, the cost and technical problems of placing high voltage lines underground are insuperable. Underground lines in the lower-voltage distribution network are feasible, but they cost from nine to sixteen times as much as overhead lines in suburban areas, and up to forty times as much in rural areas.

The power grid connecting load centers and generating centers has become bewilderingly complex, to assure reliable service and ability to meet peak or unexpected demands and emergencies. The interconnection of power systems was stimulated by the shocking blackout in the northeast in 1965. Already, a 500-kilovolt loop spans a major section of the eastern seaboard from north to south, and strengthening of east-west connections is underway. Major load centers in Michigan, Indiana, and Ohio are connected, forming part of a loop around Lake Erie to connect centers in Canada and the United States. Construction of lines connecting the Pacific northwest with the Pacific southwest is in progress. Interconnections between utilities in the Great Plains and the Rocky Mountain areas are more difficult because the load centers are smaller and farther apart, but there are plans for development on a modest scale. As the transmission system grows in size and complexity, the location of a power plant will become independent from the location of the need for power.

Whether this grid is necessary can be debated. The cost in dollars will be about $8 billion, in the period from 1967 through 1975. The environmental cost has not been measured, nor has the net value to society. It is as sobering as it is comforting to think of this stupendous web crisscrossing the country from Atlantic to Pacific and from Canada to Mexico, able to carry power from anywhere to anywhere in seconds. Although it serves us, we are at its mercy.

It could be argued that commitment to a great web of unsightly, sputtering, humming, and crackling overhead transmission lines is premature because of potential improvements in underground transmission. Low temperature transmission holds the most interest. When some materials are cooled to very low temperatures their resistance to the passage of electricity decreases until it suddenly vanishes and they become superconductors. At first this phenomenon was merely a laboratory curiosity, but it soon became a laboratory tool. Work is now underway to develop underground transmission of electricity based on the principles of superconductivity. Table 8-3 compares some properties of several proposed low temperature systems. The hydrogen-cooled system is of most interest, because of the large gain in capacity per degree of cooling. A superconducting transmission line sounds very attractive, but the problems of cooling thousands of miles of underground cable to 7 degrees above absolute zero may be knottier than even conservative technologists expect.

In fact, the practicality of any low temperature system is still open to question because no commercial system has been designed yet. Economic estimates are unreliable because, as in the case of fusion reactors, no one is certain what is being estimated. The cost in both dollars and energy of

Table 8-3. Operating Temperatures and Relative Load Capacities of Low Temperature and Superconducting Transmission Systems

Coolant	Operating Temperature F°	Relative Conductivity of Metal Transmission Line
Oil (reference)	+68	1
Liquid nitrogen	-322	8
Liquid hydrogen	-422	500
Liquid helium	-452	superconductor

refrigerating the transmission lines may be so great that the system is rendered useless.

The supply of electricity must respond almost instantaneously to variations in demand. There are daily, weekly, and seasonal cycles of demand, which vary with the major uses of electricity and the climate. A persistent problem in the electric power industry has been to find ways of meeting peak demands and of using the generating capacity when demand is low. Systems operate most efficiently and economically at full load, and for this reason, despite current shortages of electricity and fuel, the power companies still encourage off-peak use of power.

East-west connection of power systems will permit exchange of power between time zones, as the time of peak demand marches daily across the country. Traditionally, additional generators have been brought into service when extra power is needed. A hydroelectric plant is ideal for this purpose because it can be started and stopped quickly. The flexibility of hydropower spurred development of pumped storage systems. Here, surplus power generated during off-peak hours is used to pump water from a low reservoir to a higher one. When demand is high, the water is allowed to flow back to the lower level, and its kinetic energy is converted to electricity as in any hydroelectric plant. But because of the laws of thermodynamics, more energy is expended in pumping the water to the higher level than is recovered when it falls. Therefore pumped storage is not a source of energy, but merely a technique for balancing temporal variations in supply and demand.

Gas turbines and diesel engines are also widely used to meet peak demands or to supply power in emergencies. Both can be started and brought to full capacity within a few minutes. Both have negligible requirements for water, which renders their siting independent of water supply. The units are prefabricated and can be put into operation in a few months — a major advantage in these days of power shortages and delays in construction of major new facilities.

A gas turbine, as the name suggests, is driven by hot combustion gases rather than by steam or water. Most gas turbines burn clean fuel — natural gas or the lighter petroleum fractions. They are noisy and inefficient, but their advantages outweigh these disadvantages for the purposes they usually serve.

Diesel engines are widely used in transportation, in mining and the petroleum industry, for generating electricity in isolated areas, and for supplying power in emergencies. They are designed to burn either natural gas or various petroleum fractions, and are more efficient than either gas turbines or small steam turbines. Their economy, versatility, and mobility make them desirable additions to utility systems even though their capacity is limited to about 6 megawatts. Total installed diesel capacity in the electric industry is expected to double between 1970 and 1980, from 4000 to 8000 megawatts, and to triple between 1970 and 1990.

The efficiency of steam electric plants is limited by the temperature range within which they operate. Because the lower temperature is set by the temperature of the environmental heat sink, efforts to increase the temperature range must be directed toward increasing the upper temperature. Much higher temperatures can be achieved in burning fossil fuels than can be used in a steam system, where the materials cannot tolerate more drastic conditions than those already imposed on them. The practical limit is about 1050° F and 3500 pounds pressure per square inch.

In the past there was some effort to find an alternative vapor to steam, which could take advantage of the high temperatures at a lower pressure. Some plants were designed to work on a mercury vapor-liquid mercury cycle, but they were not satisfactory. There is presently no other candidate for an alternative vapor.

There are other ways, however, to take advantage of the temperatures that combustion of fossil fuels can provide. When used in conjunction with a steam cycle, they are called topping cycles (and the steam cycle, appropriately, is the bottoming cycle). Waste heat from the topping cycle heats steam for the steam cycle, thus increasing the temperature range in which the combined system operates and raising its efficiency. Optimistic estimates say the efficiency of some combined cycles could reach 70 percent, but since these advanced systems have yet to be put into operation, there are still many unknowns.

Most promising for the near future is a topping unit that uses hot combustion gases directly to drive a gas turbine, with the somewhat cooled gases then rejecting their heat to a conventional steam generating system. Some gas turbines already operate in this way to provide extra power for peak demands, although most of them operate without the steam cycle. This system, and gas turbines in general, must burn clean fuel in order to hold air pollution and maintenance costs to tolerable levels. Current research emphasizes a gas turbine topping cycle in conjunction with gasification of coal for onsite production of electricity.

The field of magnetohydrodynamics (MHD) holds promise for greater efficiency in energy conversion, although difficult problems with materials and design block the road between theory and application. MHD is based on the fact that an ionized gas, passing through a magnetic field, generates a current of electricity. In an MHD generator, a stream of hot gas takes the place of the cumbersome steam turbogenerator, allowing temperatures up to 5000° F in an efficient, one stage conversion process.

Plans call for fossil fuels to be burned in pre-heated air in a high-temperature combustion chamber. The combustion gases will become highly ionized, and the electron density will be further increased by seeding the gas with a readily ionized salt like potassium sulfate. Ionized gas and seed will then be released at high velocity through an MHD channel, across an intense magnetic field. As they pass through the channel they will induce a direct current voltage between electrodes in the channel walls. The MHD cycle can be used either alone or as a topping unit for a gas turbine or steam cycle. At somewhat lower temperatures, liquid potassium, cesium, or sodium could replace the hot gas as the ionized fluid in the MHD channel. This alkali metal system might use heat from a fission reactor or from the combustion process.

One problem common to both gas turbine and MHD cycles is that, at the high combustion temperatures involved, much larger amounts of nitrogen oxides are formed than are formed in the conventional fossil fuel plants of today. At present there is no satisfactory way to decrease their formation or remove them from stack gases.

Two other methods of generating electricity convert heat directly to an electrical current, bypassing the intermediate conversion to mechanical energy. One of these is called thermionic generation and is based on the thermionic effect — the phenomenon of electron emission, or the boiling off of electrons when a metal is heated beyond a certain point. The process requires temperatures above 2500° F, and at such temperatures the functional life of the components is short. Although thermionic generation has useful applications in space technology, large-scale commercial applications seem unlikely. A similar situation holds for thermoelectric generation, which is based on the voltage difference produced between them when two different conductors are joined and heated. Thermoelectric generation powered by the sun or another heat source has potential applications in communication and other specialized areas, but like thermionic generation it is very inefficient and has little to recommend it for large-scale power generation.

When we say that a power plant generates electricity with an efficiency of 40 percent we are looking at only a small segment of the chain of events that begins in the coal or uranium mine or the oilfield and ends when you switch on an electric light in your home. For each 100 kilowatt-hours of energy in the form of fuel in the ground, about 35 are lost during extraction or otherwise rendered irrecoverable. Perhaps 2 more kilowatt-hours are lost during processing and transportation. Thirty-nine are wasted in the conversion to electricity. Two are lost in the transmission of electricity to the consumer. The final conversion may be very efficient or very

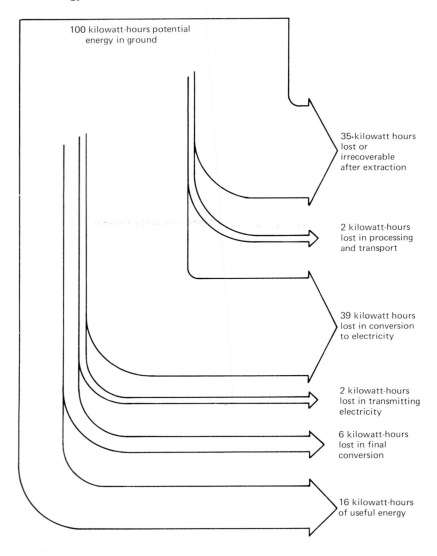

100 kilowatt-hours potential
energy in ground

35 kilowatt hours
lost or
irrecoverable
after extraction

2 kilowatt-hours
lost in processing
and transport

39 kilowatt hours
lost in conversion
to electricity

2 kilowatt-hours
lost in transmitting
electricity

6 kilowatt-hours
lost in final
conversion

16 kilowatt-hours
of useful energy

fig 41

Figure 8-2.   Energy losses in production and use of electricity. Starting from fuel in the ground, the fate of energy to final use is traced.

inefficient, but we will assume an average efficiency of about 75 percent, which means that 6 more kilowatt-hours are lost. Thus only 16 of the 100 kilowatt-hours of potential energy contained in the original fuel are actually converted to useful work — an overall efficiency of 16 percent (figure 8-2).

But even now the story is not complete. At each step of the way a great deal of energy is required to maintain the energy system. The refining of petroleum uses 10 percent of all energy used by industry. The fabrication of uranium fuel rods requires large amounts of energy. Gas pipelines use about 4 percent of the energy used by the transportation sector of the economy. Rail and marine transport of fuel also requires energy. Up to 10 percent of the output of a power plant can be consumed onsite in devices for pollution control. Beyond that, there is a substantial investment of energy in all the equipment that brings energy to us in the final form of electricity. Perhaps, after all, we should not feel so superior in our efficiency to a green plant.

References

1. F. Warren, *Electric Power and Thermal Discharges*, ed. M. Eisenbud and G. Gleason (New York: Gordon and Breach, Inc., 1969), p. 24.
2. G. Garvey, *Energy, Ecology, Economy* (New York: W. W. Norton & Company, Inc., 1972), pp. 157-173.

PART THREE

PROSPECTS FOR THE FUTURE

9

WHERE DOES IT ALL GO?

*In what we call progress, 90 per cent of our efforts go into finding a
cure for the harms linked to the advantages brought about by the
remaining 10 per cent.*

Claude Lévi-Strauss

When passing Gary, Indiana or Pittsburgh, Pennsylvania, it is easy enough
to identify the miles of factories with energy use. But it is not always so obvious just where the energy goes, or what it is used for.

Americans should learn something about their patterns of energy consumption. Only with such knowledge can we make intelligent choices
about changing. For in the final analysis, we have only two options in
energy use: we must either find ways to supply ever increasing amounts of
energy, or curb our demands.

One way to meet the energy crisis is to slow the growth in consumption. There are some clear benefits to be gained from slowing down. It
would give us more time to supply the energy that is needed and would
also conserve our supply of fuels for needs still further in the future. And
for each gallon of fuel not consumed, there is less pollution to control.

For many people, the growth in energy use is an abstract statistic, unrelated to their own lives. Most people have very little idea of the amount
of energy they use. A questionnaire on energy consumption was prepared
by a group of students at the University of Wisconsin and administered to
more than one thousand people. The students could not even find a unit or
a measure of energy that people knew well enough to answer a question.
Watts and calories sounded familiar to many people, but they could not

relate these units to their own domestic energy use. Kilowatt-hours were even less useful, while joules and ergs brought only perplexed stares. Finally settling on the 100-watt light bulb as a unit, the students found that most of those questioned could not come within 10,000 percent of guessing their own energy use.

So again we ask: where does it all go?

Conventional sectoral breakdowns of energy use have been dictated in part by the way in which statistics are collected, in part by the needs of government economists, and partly by historical accident. Figure 9-1 presents one of the most common sectoral divisions of energy use. While there are a number of difficulties in this kind of partitioning, figure 9-1 clearly indicates that the division of energy consumption among these sectors has been remarkably stable for the last twenty-five years. During this period energy use has doubled, yet the only trend observable is gradual enlargement of use in the sector labeled "other."

The industrial sector is the most homogeneous; more than 70 percent of the elements included in this sector would correspond with our usual

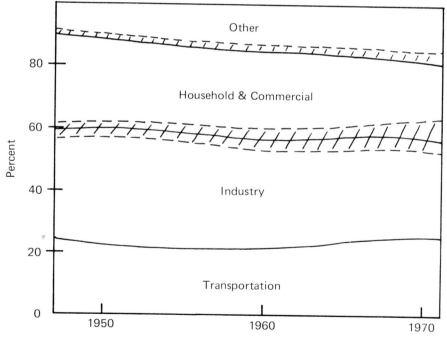

Figure 9-1. A traditional sectoral division of energy use, 1947-1971. (The shaded area represents energy consumed for the generation of electricity.) (Adapted from E. Cook, "The Flow of Energy in an Industrial Society," *Scientific American* 224, no. 3 [1971]: 137.)

idea of manufacturing industries. Among the six industries which account for two-thirds of the energy used by this sector are food processing and petroleum. Food processing accounts for more than 5 percent of all industrial energy use (chapter 4). In 1968, the petroleum industry and related fuel producers accounted for about 11 percent of industrial use. However, only one-third of the fuels produced are consumed by industry. If the energy costs of the petroleum industry were assigned to the ultimate fuel users we might better understand the savings to be gained by measures to reduce energy consumption. In figure 9-2 we have a breakdown of energy consumption in the industrial sector. A more detailed look at the six leading industrial energy consumers will be useful.

Two-thirds of the energy consumed in production of primary

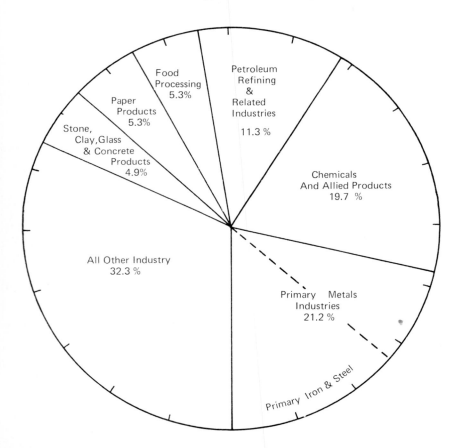

Figure 9-2.   Division of energy use in the industrial sector. Source: U.S. Office of Science and Technology, Executive Office of the President, *Patterns of Energy Consumption in the United States* (Washington, D.C.: 1972.)

metals goes to the basic iron and steel industry. During the 1960s energy use was reduced by about 10 percent per unit of output, primarily because of the more efficient use of blast furnaces.

Statistical data suggest an even larger decline in energy consumption because they consider only the amount of energy actually used at the site of steel production. But use of electricity has increased by about 20 percent in response to several factors, including the need to meet air and water quality standards. If purchased electric power is charged to the steel industry in terms of primary sources of energy required for its generation, the decline in energy use per unit output is near 10 percent rather than the higher figures sometimes quoted. Electric energy accounted for about 10 percent of the energy used for iron and steel production in 1969.

The use of natural gas in the iron and steel industry doubled during the 1960s. The price increases associated with the shortages of domestic natural gas mean that this trend cannot continue. Use of natural gas in the steel industry will probably decline (or at most remain stable) for the remainder of this century. This change, together with the need to meet environmental standards, will increase the use of electricity in steel production.

Most experts suggest that further efficiencies are possible, but savings in energy per unit output have been decreasing (four-fifths of the decline in energy use per ton of product were obtained between 1960 and 1965 and only one-fifth between 1965 and 1970). Further gains in efficiency will probably be balanced by the increased use of electricity and the additional energy needed to meet environmental standards during the 1970s. Further decline in energy cost per ton of iron and steel appears unlikely after 1980. If the demand for more exotic alloys of steel continues to increase, the energy cost of steel may start to climb.

For the remainder of the primary metals industry, the trends in energy consumption are not promising. About 12 percent of the energy used in the production of primary metals was consumed in the production and processing of aluminum, while 8 percent went to ferrous foundries, 4 percent to copper production, and the rest was scattered among zinc, lead, and the processing of metals. Aluminum production increased by 88 percent in the 1960s, and aluminum is by far the most energy-consumptive of this group. The energy needed to process recycled metals is much less than the energy required to produce virgin metal. This is especially true for aluminum and copper, as the primary production processes are mainly electrical.

Except for zinc, recovery of the principal metals improved during the 1960s. The most dramatic improvement in recovery was a whopping 160 percent increase in recovered aluminum. By the end of the decade recovery rates were almost 20 percent of primary production. Despite this and some other efficiency-based reductions in energy use per unit output, future energy use will almost certainly increase as aluminum production expands in response to shortages and the high cost of copper and the base metals.

In this industry a growing proportion of energy is used for rolling, drawing, casting, and forging of metals. This trend appears to be part of a long-range development in technology, and will continue. Thus, gains in efficiency in foundries are likely to be canceled by the need for more and more complicated (and energy-consumptive) processing.

New air and water quality standards mean higher energy consumption per unit output. It will take another decade of experience to learn how much energy will be needed to meet these standards.

The second largest industrial user is the chemical industry. In this area uses for energy are as diverse as the estimated ten thousand industrial chemicals manufactured. Even the industry is not very well defined. There are some giant chemical producers (names like DuPont, Dow, Union Carbide, and Olin come to mind), and the largest twenty companies account for three-fourths of production in most categories. Nevertheless, 43 percent of the five hundred largest corporations manufacture chemicals, often as byproducts of their primary product. The trend from coal to oil and natural gas as the primary raw materials of chemical production in the past thirty-five years has been followed by entry of the oil industry into production of chemicals

Fuel use in the chemical industry differs notably from that in other industries, for about half the fuel consumed is used as raw material rather than for production of energy. This use has been growing rapidly. Use of oil and natural gas as the basis of chemical production is less sensitive to higher prices than are energy uses, and thus it will probably continue to expand, even at the expense of energy uses. Increased production of coal, or of gasified coal, may supply some of this expansion.

Two-thirds of the energy used in the chemical sector goes to the production of eleven products: chlorine-caustic soda, synthetic soda ash, acetylene, oxygen, carbon black, ammonia, manufactured aluminum oxide, phosphorus, methyl alcohol, sulfur, and ethylene and related products (including propylene, butadiene, and aromatics). Most of these products are supplied to other industries as materials for manufacture of final products. In most cases we must look to the final products to determine whether savings are possible. Some means of producing these eleven products are energy-intensive and others are not (table 9-7). Often there are several choices available for the manufacture of chemicals. These choices may differ in energy requirements per unit output by 1000 percent or more. Which is chosen depends upon several factors, but prominent among these is the absolute price of energy and the relative price of energy and labor. The future of energy use in the chemical industry depends also on the degree to which substitution of manufactured for natural materials continues. For example, many products of the chemical industry have replaced textiles, wood, hides, rubber, and mined chemicals. Production of chemical foods is possible, although insignificant at present.

Petroleum refining and associated industries are the third largest industrial consumer. However, it might be desirable to apportion these energy costs among the fuel users. It takes about 200,000 kilocalories per barrel to process crude oil into the more than one thousand finished petroleum products on the market today. This amounts to about 11 percent of the energy content of the oil. The proportion of energy used per barrel of crude oil has changed little in the past ten or fifteen years, and no great changes in this energy consumption are expected in the near future. Newer refineries succeed in obtaining a more profitable (and, at times, more useful) mix of products, but do not achieve any significant decrease in energy consumption per unit output.

Although food processing has reduced its percentage share of total industrial energy use, its absolute increase for the past decade has exceeded increases in output. Processed food accounts for an ever larger portion of our diets, although the return to natural foods could reverse this trend. Decreasing the energy consumed in food processing will be difficult, because the four largest energy consumers in this area (meat, dairy products, grains, and sugar) are not much affected by the change in lifestyles (with the possible exception of some reduction in meat consumption).

Consumption of paper has increased enormously as our society has grown more complex. Use of forms, reports, packaging, and packaging within the packaging seems to expand each year. Energy use in the paper and paper products industry has grown by about 20 percent in the past twenty years. This expansion in energy use has been outstripped by the increase in production, and energy use per unit output diminished by about 15 percent during the 1960s. This decline is due partly to changes in the major types of paper products (for example, production of kraft paper, used for packaging material, requires less energy than production of other types), and partly to technological improvements in the manufacturing process. As these improvements spread through the industry, some further reduction in energy per pound of paper may be possible, but improvements will be harder to come by in the future. Pulp yields, for example, increased from 96 to 98 percent between 1958 and 1967. The next improvement will be much more difficult, and we will never reach 100 percent recovery.

Many of the raw materials for stone, clay, glass, and concrete products are mined or they are obtained from the byproducts and waste of other industries (slag and sludge, for example). Most of the energy consumption is for mechanical crushing, grinding, and blending and for heat to obtain chemical changes at high temperatures. In 1967, cement production con-

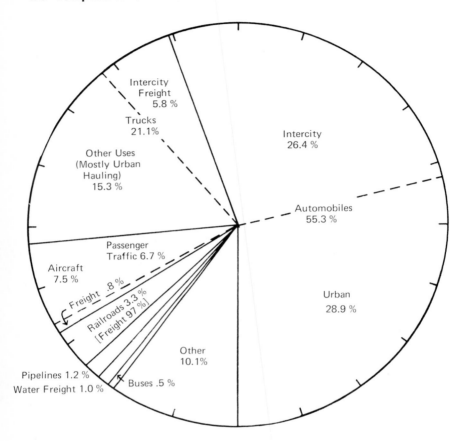

Figure 9-3. Division of energy use in the transportation sector, 1970. ("Other" includes general aviation, non-bus mass transit, recreational and passenger boating, recreational vehicles, and other uses.) Source: E. Hirst, *Energy Consumption for Transportation in the United States* (Oak Ridge National Laboratory Report ONRL-NSF-EP-15).

sumed 43 percent of the energy in this group, glass production consumed 16 percent (the largest end use is glass containers), concrete products consumed 15 percent, and structural clay products 14 percent.

Thus, cement, concrete products, and structural clay products together account for almost three-fourths of the energy consumed in this industry. They are used primarily in the construction of roads, buildings, and the like. A functional look at energy consumption would assign their cost to the construction industry, and the next largest item — glass containers — would be assigned to the food system. Energy costs of these products vary considerably according to the kiln time and temperatures needed. While there have been improvements in kiln design and certain economies of scale are possible, there are few prospects for dramatic energy savings per unit output in this sector. Even recycling offers little help here.[1]

Use of energy by the transportation sector has undergone major changes, as have the principal modes of transportation. In years gone by, the transportation services moved both goods and people by the same methods. Currently, however, those modes of transport which handle most freight carry very few passengers, and vice versa. Figure 9-3 presents a breakdown of energy consumption of the transportation sector. This is the only sector in which the use of electricity is not expanding. Even though electric mass transit is attractive in many ways, this mode of transportation was never used extensively in the United States, and it has declined in recent years. Most transportation is powered by petroleum products, and 53 percent of the total refinery output goes to this sector.

The transport of people is dominated by the private automobile. In 1970, autos accounted for almost 88 percent of all intercity passenger miles, a proportion that has seen little change in the past twenty years although the fuel efficiency has worsened. Air travel, the least efficient mode of passenger travel, accounted for 9 percent of intercity passenger miles in 1970 — up from 4 percent in 1950. Here too, there has been declining efficiency of fuel usage. Air transport has used almost one-third more energy per passenger mile in the past decade. Most of the remaining intercity passenger travel is provided by buses, although as recently as twenty years ago, railroads provided more than 5 percent of all intercity passenger miles. Increased fuel consumption per passenger mile in aircraft and automobiles is not primarily due to technological developments. For aircraft, scheduling has resulted in a drop in percentage load and a corresponding increase in energy cost per passenger mile. The new, larger jets contribute a share to the increase. For autos, the increase is the result of ever larger engines and the growing number of energy-consuming options, from air conditioners to devices for pollution control.

Table 9-1 summarizes the modal split and the efficiencies of intercity

Table 9-1.   Intercity Passenger Transport, 1970

Mode	Billions of Passenger Miles	Percentage of Total Passenger Miles	Energy Consumed (trillions of kilocalories)	Efficiency (kilocalories per Passenger Mile)
Automobiles	1020	88.1	940	920
Airplanes	104	9.0	306	2940
Buses	25	2.2	7.6	308
Railroads	8	0.7	1.9	240

Sources: *Statistical Abstracts of the United States;* Interstate Commerce Commission; Federal Aviation Administration; Department of Transportation gross data.

The largest single end use for energy in the United States is the private automobile. In earlier days cars were simpler, more functional machines and far
less numerous. Today, many urban problems are traceable to the automobile. Waste disposal and waste of resources are often auto-related
problems. Yet road building continues as if we were intent on paving over
everything. (Left photos courtesy of Marine Studies Center, University of
Wisconsin. Upper right photo by Sandy Levitz. Lower right photo by
Marshall Henrichs.)

passenger transport in 1970. This choice of modes was not made with energy use, or even minimum cost in dollars, in mind. Energy consumption per passenger mile has been rising steadily for twenty years or more. It seems doubtful that this trend can continue.

Over half of the total fuel energy for transportation is consumed in urban areas. Since we expend this energy in less than 2 percent of the land area, and the remaining half in the other 98 percent of the nation's area, our urban air pollution problems should not surprise us. The auto dominates intracity transport, with 94 percent of the total passenger miles. Well over half of all auto trips are in urban areas and for less than 5 miles. Table 9-2 summarizes the data for urban passenger transport in 1970. Although there are no data available for walking and bicycling, anecdotal evidence suggests that both have been declining for many years. The recent rise in bicycle sales may reverse the trend for bike travel, but this increase only began in the 1970s.

There has been a dramatic decline in mass transit: from 31 percent of urban passenger miles in 1950, mass transit dropped to 13 percent in 1960, and is still falling. It is surprising that the efficiency has not also declined precipitously. Note that the efficiencies given in table 9-2 are from experiential data, thus they differ from the usual estimates. Before concluding that we need only walk or ride bicycles to solve our energy problem, we should recall the discussion of energy in the food system (chapter 4). The energy required for walking and bicycling is derived from our food system, and at the present energy cost of a kilocalorie of food, it is more efficient, in terms of energy conservation, to take the bus.

To a considerable extent, the nature of a freight shipment determines the mode of transport. It is foolish to consider shipping iron ore by airplane, and it is quite impossible to ship strawberries by pipeline. To a large extent, then, the present modal mix of freight shipments is determined by what we have to ship. There have been changes in the proportions of

Table 9-2.   Intracity Passenger Transport, 1970

Mode	Billions of Passenger Miles	Percentage of Total Passenger Miles	Energy Consumed (trillions of kilocalories)	Efficiency (kilocalories per Passenger Mile)
Automobile	739	94	1130	1530
Mass Transit (almost all bus)	49.5	6	14.8	300
Walking	—*	—	—	75
Bicycle	—	—	—	45

* — = no data
Sources: *Statistical Abstracts of the United States;* Department of Transportation; Interstate Commerce Commission; and Federal Aviation Administration gross data.

freight carried by various modes in recent years. In part these shifts are due to changes in industrial activity and in part they represent real choices in competing modes — usually for reasons of economy or convenience. The most notable of these shifts has been the expansion in freight transport by truck at the expense of rail freight transport. For their specialized uses, pipelines have also cut into rail freight owing to the lower cost per unit shipped. The railroads' share of freight transport has dropped from more than 60 percent of the total in 1940 to 40 percent in 1970. The shift to trucks is further explained by the superhighways which have been built at public expense. These provide an indirect subsidy to trucks as a mode of transport. Table 9-3 presents the modal division of freight transport in 1970, together with the associated energy costs.

Present trends suggest a continuing shift from rail to truck if, and there are two big ifs, the construction of high speed roads continues and the price of energy does not rise too dramatically. Air freight, which has expanded its volume tenfold in the last twenty years and doubled twice in the past ten years, must slow its rate of growth. Should the same growth continue for another twenty years, air freight would carry 2 or 3 percent of the total freight and consume almost half of the freight transportation energy. This is unlikely to say the least. Efficiencies of pipelines, water carriers, and railroads have been improving and could be improved further. Trucks seem not to be improving their efficiency much.

Energy used by the residential sector has declined slightly as a percentage of the total, but the actual amount used is increasing steadily. "You've come a long way, baby" begins one particularly obnoxious cigarette advertisement. In the American home, the changes in lifestyle in the last two generations are staggering. One evidence of the change is the large number of appliances to be found in most homes. There is a tendency to confuse this increase in number of appliances with increased energy use, but such is not the case. The bulk of residential energy use still goes

Table 9-3.   Freight Transport, 1970

Mode	Volume (billions of ton miles)	Percentage of Total	Energy Consumed (trillions of kilocalories)	Efficiency (kcal per ton-mile)
Railroads	768	40	131	170
Trucks	412	21.4	370	900
Pipelines	431	22.4	49	114
Water carriers	307	16	42	136
Airways	3.4	0.2	34	10,000

Sources: *Statistical Abstracts of the United States;* Interstate Commerce Commission; Department of Transportation; Federal Aviation Administration gross data.

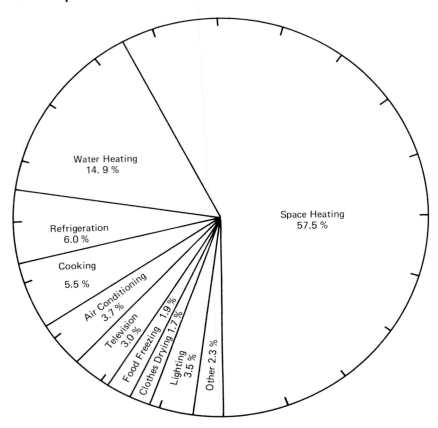

Figure 9-4. Division of energy use in the residential sector, 1968. Source: Office of Science and Technology, Executive Office of the President, *Patterns of Energy Consumption in the United States* (Washington, D. C.: 1972).

for space heating and for storage and preparation of food. Meanwhile, our favorite candidates for foolish energy use, the electric comb and the electric knife, consume less than one-ten-thousandth of residential energy.

Figure 9-4 shows the distribution of residential energy consumption in 1968. Heating is the big consumer. Although heating declined as a percentage of total residential energy consumption in the last decade, the average energy expenditure for heating per household rose by almost 20 percent. Any examination of future prospects for energy use must look closely at trends in home heating. The fastest growth in residential energy use is for electric heating. Between 1960 and 1968 energy used for this purpose increased fivefold, but even in 1968 only 5.6 percent of all residences were heated electrically, and they used less than 7 percent of all energy used for residential space heating. Because of the low efficiency of fuel use in electric heat (see chapter 3), growth in the use of electric heat could increase domestic energy requirements enormously. About one-third of the increase in per household heating between 1960 and 1968 is due to the growing use of electric heat. The remainder is not easy to account for. The most likely explanation is that the general rise in affluence, including larger housing space, has caused the increase. Between 1960 and 1968 the annual consumption for electric heat per household increased from 11,908 kilowatt-hours to 14,153 kilowatt-hours — an increase of about 20 percent. We would like to stress this point because many believe that increases in residential energy consumption are primarily due to the number and size of large appliances used.

The second largest residential use is water heating. In 1960, 75 percent of all residences had water heaters, but by 1968, 94 percent had them. In addition, the energy consumption per unit rose by a little over 5 percent. In this case the increased consumption is probably due to the growing number of dishwashers and automatic washing machines.

Refrigeration and cooking consume about equal amounts of energy, and both are chargeable to the way in which we feed ourselves. But there the similarity ends. Energy consumption for ranges declined 3 or 4 percent during the 1960s, despite the addition of such features as self-cleaning ovens, while energy consumption for refrigerators increased more than 50 percent. The increase for refrigerators is a consequence of the growing popularity of frostless units, which require about 50 percent more energy. Since the refrigerator runs continuously, added features are reflected in increased energy costs. The average storage capacity of units has increased in recent years, more than canceling some improvements in insulation technology. The decline in energy consumption for cooking is almost certainly a reflection of changes in cooking and eating habits, such as the trend toward more pre-cooked and instant foods. Since such pre-cooking and processing is then charged to the food processing industry, the total energy bill for food preparation probably has not declined.

Other residential uses do not constitute large segments of domestic energy consumption, but of these, air conditioning and clothes drying deserve special mention. Between 1960 and 1968, the average annual growth rate for residential air conditioning was 16 percent, which implies that energy use for air conditioning increased more than threefold in that period. At the same time, the consumption per unit for room-sized units increased by 10 percent. If more than two room-sized air conditioners are wanted, there is an apparent energy economy to be had by using central air conditioning. Often the actual result does not bear out the energy saving, however, because of the tendency to turn on the central unit at the outset of warm weather and allow it to run in response to a thermostat, whether needed or not. Most new homes are designed for central air conditioning. This represents a commitment to use of air conditioning for the life of the house, since a house so designed is difficult to cool by any other method (such as fans or open window ventilation patterns).

The use of clothes dryers is expanding. In 1960, 19 percent of homes had clothes dryers; by 1969, 40 percent had them. For electric units, per unit consumption has risen slightly, while the reverse has happened for gas units. It is difficult to estimate the level of saturation for these appliances. Again a lifestyle change is involved. It is clear to anyone touring a city neighborhood of even very modest means, that outdoor clothes drying has almost disappeared, even in warm sunny weather. Here is a clear example of the way technology changes our lives, in ways not anticipated by the individual. An item of undeniable convenience in winter (ask any mother with a diaper-laden wash) has become the norm even when not necessary or particularly more convenient. And to restore something lost from the sunlight outside, some dryers will provide imitation sunlight with the process.

In the center of any middle-sized American town there is a profligate use of energy that is scarcely matched in any other country. As evening comes, the signs light up and the garish kaleidoscopes of blinking colors compete for one's attention. Mercury vapor lamps hanging from steel poles make one's companions look like corpses. A commercial strip, we say, and hurry through to our homes or jobs.

The commercial and "other" sectors include many activities that are not so obtrusive — health services, motion pictures, wholesale and retail trade, offices, schools, museums, commercial farms, fisheries, hotels, motels, and a host of minor pursuits. The inclusion of agriculture in this sector emphasizes the unsatisfactory nature of the present sectoral divisions. All governmental activities are included in the "other" sector.

In figure 9-5 we attempt a functional division of energy use for the commercial and "other" sector. Data for 1960 were used for this sectoral subdivision because more recent data showing defense consumption were unavailable. More energy is used in defense than in all other government activity. The remainder of government use was scattered and surprisingly small, in view of the fact that one worker in ten was employed by the government in 1960. The rise in government employment — in 1970, 15 percent of all employed persons worked for the government — suggests that this use is growing, but it is still not especially significant. Government is primarily a service activity and, as such, is a low energy consumer. In any case, the total amounts involved suggest that measures like turning off the lights in government offices are not likely to alleviate energy shortages, although they may be of symbolic value. Space heating, asphalt and road oils, air conditioning, outdoor lighting, and perhaps water heating almost certainly account for more than half of the non-defense government usage.

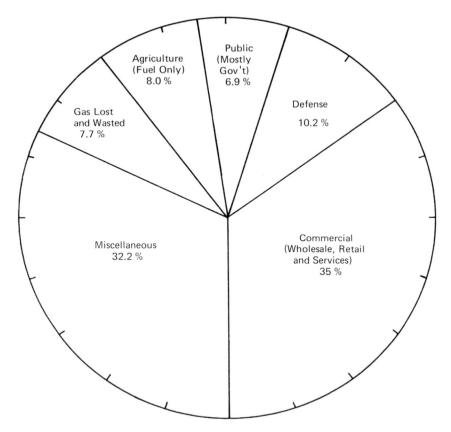

Figure 9-5. Division of energy use in the commercial and "other" sector, 1960. Source: H. H. Landsberg and S. H. Schurr, *Energy in the United States: Sources, Uses, and Policy Issues* (New York: Random House, 1968).

In the commercial sector, end uses divide into categories much like those in the residential grouping. This is not surprising when it is remembered that the sector includes offices, hotels, restaurants, and large apartment buildings. The most rapid increases in the commercial sector are in air conditioning (about 90 percent of new commercial space is air conditioned) and in "other," which is probably a reflection of the increasing mechanization of office work, quick food service and the like. The energy consumed by air conditioning doubled between 1960 and 1968, while "other" use tripled. Such growth rates cannot be sustained very long.

Our increasing affluence and resulting change in habits have produced other developments in the commercial sector. Commercial establishments served 17.5 percent of all meals in 1968, compared to 15 percent in 1960. Not only are more people eating out, but more live in universities and other residential institutions. In general, commercial establishments have been more economical with energy than the home. Despite serving an increasing proportion of meals, commercial establishments increased their consumption of energy for refrigeration and cooking more slowly than did the residential sector.

There are many ways of classifying energy consumption. Upon leaving the sectoral classification of energy use, we should review the activities which consume the greatest amounts of energy. In table 9-4 we present those activities which accounted for more than 1 percent of the energy consumed in 1970. Together, the seventeen end uses listed in table 9-4 accounted for two-thirds of all U.S. energy consumption.

Of the seventeen activities listed, five involve direct personal use of energy, six are industries which produce consumer products, two involve commercial transportation, and two are the price of energy production itself. If we look at energy consumption in this way, energy statistics lose their abstract, impersonal quality, and we may be able to relate energy consumption to our everyday life.

For most people, a bit of translation is required to convert the conventional sectoral categories of energy use into the events and environment of their everyday life. To begin a view of energy use from the personal point of view, we must start with such necessities for survival as food and clothing.

Table 9-4.   End Uses Accounting for More than 1 Percent of Total United States Energy Consumption, 1970

Use	Percentage
Private automobiles	13.8
Home heating	10.4
Primary metals industry	6.8
Chemical industry	6.3
Trucks (all uses)	5.3
Petroleum refining	3.9
Residential hot water	2.7
Defense*	2.6
Agriculture (direct fuel use only)	2.0
Natural gas lost and wasted	1.9
Air passenger traffic	1.7
Food processing industry	1.7
Paper products industry	1.7
Public (non-defense government and public institutions)	1.7
Stone, clay, glass, and concrete industry	1.6
Home refrigeration	1.1
Home cooking	1.0
TOTAL	66.2

*Note: Defense use is calculated on the basis of 1960 data.

The treatment developed in chapter 4 indicates how the food system might be handled in a new classification. Our breakdown of the food supply system is as complex as any regrouping that might be attempted, since food supply involves elements from all the conventional energy use sectors. There have been striking changes in food consumption patterns and associated energy use in the past generation, and these changes are, in general, understood. For example, the proportion of meat in the average American diet has increased by 25 percent since 1950. This is largely a matter of preference and affluence, urged onward by the low price of meat relative to other commodities. The extent of packaging and processing of food is again a matter of preference and convenience. In a subsequent section we will examine the energy savings that might be made in the food system. Here we need ask only if the diet of 1970 made you healthier or happier than the diet of 1960 or 1950.

Clothing production does not consume a large amount of energy. The entire textile industry accounts for only 0.6 percent of energy use, and this industry makes a great many things in addition to clothes. Even if we make generous allowances for the supplying industries, for wholesale and retail trade, and for domestic energy use in washing and drying of clothes, we will not find clothing to be an area of high energy consumption. But even here savings would be possible without a substantial change in lifestyle. For example, production of synthetic fibers consumes more than twice the energy needed for production of the natural fibers they replace.

Shelter is a more complex matter. In a functional classification, the housing category would include the energy consumed in construction and in the building materials industry as well as the energy used in operating a home (such as space heating, hot water heating, and lighting).

Employment, and the buying power it brings, is necessary in all present social systems. Unfortunately, it does not lend itself to energy use classification and one must be content with attention to the service or product with which the employment is associated. To that end one can resort to a classification of producer goods industries, consumer goods industries, and service industries. Nowhere among the seventeen leading energy users is an activity that would be called a service industry. But a comparison of absolute energy use per job in service industries against that in durable goods industries might shed some valuable light on our societal choices. A society emphasizing service industries would consume less energy. Should we consciously aim for such a society?

Consumer goods are those products which are used directly as they are finished by the industry. They might be further divided into necessities and luxuries. To do so implies a value choice that is more an individual than a public matter. We might choose to include among necessities those products used or purchased by more than half the population. Such a division would yield data that might advance understanding of society's energy use. However arbitrarily the division is made, the energy invested in the necessities is of more importance to society than the energy used on luxuries. A rough average of energy consumption per job could be calculated for these two kinds of industry.

Producer-products industries are very large energy users. From a functional point of view it would be desirable to apportion these energy costs among the manufacturers and the consumers. Such a division is beyond elementary techniques, but well within the methods devised by Leontieff and others for tracing economic flow. Apportionment of energy costs among end products would make clear the energy requirements of goods and the energy requirements per job. It seems that basic statistical data to accomplish this kind of classification are available.

In a functional analysis of energy use, the conventional transportation sector would be divided into two parts. The energy used in the transportation of goods would be charged to the goods transported. All that would remain is the transportation of people. This area subdivides easily into necessary transport associated with one's employment, and transportation for leisure or recreation. Such a division would be advantageous because, for most of us, the issues in work-related transportation are convenience, economy, and speed rather than preference for a given mode of transport. Transportation for personal or family business (shopping, medical care, banking, and so on) should be included under necessary transportation. Transportation for employment and family business account for two-thirds of all automobile use, and more than 60 percent of these trips are less than 5 miles.[2]

There are some uses of energy that are not easy to relate to individuals. Public uses such as street lighting, government services, defense, education, and the like are part of the common effort to make individual life more pleasant and free from threat. These uses should retain a category of their own. The energy consumption and losses associated with the procurement, conversion, and transport of energy supplies could be assigned in proportion to fuel and energy used, but it might be more informative to maintain a separate category for these. If we had such a category at present, it would be clear that the Atomic Energy Commission's nuclear energy program has each year consumed more energy than it has produced.

A final category would be energy use for recreation and leisure. This category, consisting mostly of services and transportation, is small but not trivial (one-third of automobile travel would be here — 4.5 percent of all energy use). If the trend toward greater fuel use in recreation continues, this sector could become large.

If we wish to answer the question "Where does it all go?", we will be more likely to find our answer with a functional classification of energy use

Recreational use of energy, though still not a major use is growing rapidly. As the internal combustion engine becomes part of more and more vacations, fuel use rises and the quiet of the country has new noises in it. Even getting there uses more energy, because a car with a trailer usually gets poor gas mileage. (Photos courtesy of Marine Studies Center, University of Wisconsin.)

than with the present sectoral classification. This question is more than trivial if we believe that we have a right to participate in energy policy choices. Choices imply comparisons, and if asked today whether money should be spent on low-energy cars or whether the Alaskan pipeline should be put through, most Americans could only answer, "instead of what?" The suggested functional classification of energy use would subdivide energy consumption as follows:

A Functional Energy Consumption Classification

1. Food system energy
2. Clothing
3. Housing:
   a. construction
   b. operation
4. Consumer goods:
   a. "necessities" (with a definition according to use)
   b. luxuries
5. Transportation of people:
   a. for employment and family business
   b. for leisure and recreation
6. Public uses:
   a. defense
   b. other
7. Leisure and recreation
8. Conversion and transportation energy losses

Can we substantially reduce our consumption of energy? Can we do so without a major change in society? A large amount of energy is employed wastefully in the United States. This waste must stop. The first energy shortages of the 1970s caught most people by surprise. Energy had been considered cheap and infinitely available. Today it is neither. The changeover from waste to frugality will involve a change in habits for millions of Americans. But it is necessary. Energy prices are rising and the limits on the supply of fossil fuels can be seen.

The changes do not require a major reorientation of society, although, with time, they will produce a rather different society. For the present, we would like to know how much could be saved by "tightening the belt."

In the residential sector home heating is the largest single source of energy consumption. Anecdotal information suggests that the practice of turning down the heat at night was, until quite recently, less common than 20 years ago; and when it is done, the lowered temperature is more likely to be 65 degrees than 55 or 60. Temperature settings lowered only 1 degree bring a 3 to 4 percent reduction in fuel use. Thus, simply by lowering nighttime temperature settings by 5 to 10 degrees and reducing daytime temperatures 3 degrees (all Fahrenheit) we reduce home fuel consumption by 15 percent. Such a reduction represents a savings of slightly more than 2 percent of 1970 energy consumption. (In this discussion we will express potential savings as a percentage of 1970 energy consumption.)

In chapter 3 we noted that the efficiency of conversion of most gas and oil furnaces was 60 to 75 percent. In practice these efficiencies are

seldom reached for two reasons: first, the best efficiency figures are obtained only when the heating unit is operated at full load, and second, thin
layers of soot and other unattended items of maintenance reduce
operating efficiency rapidly. The National Bureau of Standards estimates
from field samples that average operating efficiencies range from 35 to 50
percent. To correct the first difficulty would require a change in the
philosophy and technology of design. But the second problem is easily
remedied by careful attention to maintenance.

Public appeals to make changes of this sort have been notoriously un-
successful, however. In part at least, there is a good reason. Given the present price of fuel and the costs of furnace service, it is often cheaper to ig-
nore the needed maintenance than to have it done, and it is too much trouble to set the thermostat down (or install one controlled by a clock). The
remedy that economics might suggest is to raise the price of fuel so that it
will be worthwhile to make these efforts. A 25 percent increase in the price
of fuels would allow home heating bills to remain unchanged if the
suggested savings measures were instituted — for fuel consumption would
be reduced 10 percent by better maintenance of heating units and 15 percent by lowered temperatures. Between 1969 and 1973 price increases for
oil have been more than 25 percent, but have been insufficient to
stimulate fuel-saving practices.

Better insulation would reduce fuel requirements for home heating.
Most houses are not well insulated. Optimum insulation (from a total cost
viewpoint) would reduce fuel needs by more than 40 percent. Each fuel
price increase would make it more economical to increase insulation efforts
still more. A modest fuel price increase would make the optimum insulation package provide fuel savings of 50 percent or slightly more. Changes
in insulation affect cooling as well as heating. If one adds some careful
control of ventilation to inhibit heat loss, further savings are possible. The
National Bureau of Standards estimates that present fuel use could be
more than cut in half.

Thus for home heating, fuel savings of 50 percent from better insulation and 25 percent from better furnace maintenance and lower
temperatures are possible. These measures should reduce United States
fuel consumption by 8.6 percent. These savings measures should also
apply to the commercial sector (figure 9-5). Since space heating of commercial establishments involves about 3.5 percent of United States energy consumption, an additional 1.2 percent of the total could be saved.

We replace about 2 percent of all buildings annually. In the design of
buildings there are many ways to reduce heating and cooling energy requirements. Heat is required only to replace whatever heat is lost to the
outside, and heat is lost only through the outside walls (including doors
and windows). If a building has more volume for a given area of outside

surface, less heat will be lost. Earlier builders knew this simple recipe and low, flat "ranch" houses were never constructed in the northern part of the country for this reason.

Domestic hot water heating consumes 2.7 percent of all energy, and, since much of it is done electrically, the fuel demand is quite large. In this area a technological change could be of enormous help. The use of solar water heaters on rooftops could provide all the domestic hot water necessary — even in areas as far north as Boston, Detroit, or southern Minnesota. It would probably be desirable to use a supplementary heater for the longest sequences of cloudy days and extremely cold weather (although water may be heated satisfactorily in bright sun no matter what the outside temperature). Use of a supplementary heater would hold down the high capital costs for such units by permitting the use of moderate-sized collector plates and storage tanks. Owing to their low operating costs, such solar units are already economically competitive with gas or electric water heaters even though their initial costs are high. How much the initial cost could be reduced by a mass production effort is unclear, but substantial reductions could be expected. Even if supplementary water heaters were allowed for, more than 2 percent of the nation's fuel consumption could be saved through solar water heating. (See chapter 7 for a more extended discussion of solar energy use.) Solar heating of residences with direct units on the roof is also possible and economically competitive with other methods, especially if fuel prices rise. Unfortunately, many existing houses would be difficult to refit with solar units.

Refrigeration and cooking consume more than 2 percent of the nation's fuel. Energy consumption per refrigerator continues to rise. Savings are possible here, both by design modifications and by changes in habits of use (a refrigerator operates most efficiently when fully loaded, as does a freezer). But in the absence of a widespread change in lifestyle, the most effective way to reduce the energy consumption of these and other appliances is a substantial rise in energy prices. Coupled with measures that indicate the amount of energy consumed by an appliance, a price rise could bring major savings in residential energy consumption. Such price rises are not likely to be popular, but they may be preferable to the continuing subsidies to the energy industry which help prevent them.

In both the residential and commercial sectors the fastest growing use of energy is air conditioning. Presently installed units and new ones offered in the market differ in their energy requirements by 100 percent or more for the same amount of cooling. There is no reason for this to continue; if enforceable regulations set standards for units, the expected growth due to this use would be slowed a good deal.

A number of other residential energy savings are possible. Although we suggest turning off lights, TVs, and stereos when they are not in use, these measures taken together would not equal the savings possible in home heating alone. This does not mean that small savings should not be encouraged. Efforts to produce these smaller savings will be welcome, and may best serve to encourage everyone to think about energy as something to be conserved.

The measures suggested above would reduce United States energy consumption by 12 percent. A sober government report on energy conservation concludes that less ambitious measures could save 5.3 percent of the expected consumption in 1980. A substantial price increase for fuel would bring more residential savings, perhaps substantially more.

Most of the comments made about household savings apply as well to the commercial sector. In addition, this sector has some special features. Most retail stores are air-conditioned. Many of them, especially suburban supermarkets, are large single-story buildings with flat, black roofs. Such a design requires an excessive amount of air conditioning. Small design changes, or incorporation of such establishments into multistory buildings, would reduce total air conditioning loads. The use of heat pumps for heating and air conditioning, and the use of engine-driven gas air conditioning units for large commercial establishments could reduce energy consumption for these uses by 50 percent. Lighted outdoor advertising, which seems to us about the most profligate use of energy to be found, must be charged to the commercial sector. Reliable totals of the energy consumed in lighted advertising signs are hard to find, but it would be easy to reduce the total expended by more than half and still maintain lighted identification of the establishment.

Government activities are usually included in this sector. As we have seen (table 9-4), the large end use here is in the defense program. Defense strategists pretend that energy shortages do not apply to them. This state of affairs seems very strange, because, while the energy shortage is new to most citizens, we have often been short of fuels during wars. If we maintained a defense posture requiring more men than we could count on, or more industrial output than might be marshaled, we would think it foolish, yet we do exactly that in the case of energy. The result is that our position on world issues must take into account our possible need for fuel in an emergency.

Our defense posture must be redesigned. A crude guess is that half of present defense fuel usage could be eliminated without changing the level of defense effort. That would amount to 1.3 percent of our total energy consumption.

Industry consumes a large amount of energy for an incredible variety of processes. In many sectors of industry, the amount of energy used for manufacturing and production processes has declined per unit product as a part of the never-ending effort to reduce costs and have a greater return.

For example, steel production is the largest industrial consumer of fuel. Through new processes, notably the basic oxygen process, energy consumption in steel production may be reduced 25 percent by the year 2000.[3] Manufacture of aluminum requires more than three times as much energy per ton as manufacture of steel, and titanium requires more than six times as much. Table 9-5 shows the energy requirements for manufacture of a variety of materials. (Such a table should be used with caution, however. A ton of aluminum will make twice as many cans as a ton of steel.) The source material for metals can make a large difference in the energy required for manufacture. Titanium manufactured from high grade titanium-bearing soils requires almost twice as much energy per ton as titanium manufactured from the mineral rutile, and recycled titanium scrap takes only one-third as much. A change in ore grade of copper from 1 percent copper sulfide (which is only one of several copper ores mined) to 0.3 percent copper sulfide doubles the energy required for production.

The situation is even more complex in the chemical industry. One chemical engineering text describes eleven processes in commercial use for

Table 9-5.   Energy Cost of Production of Selected Materials

Material	Energy Required (million kcal/ton)	Range of Values Cited (million kcal/ton)
Metals		
Steel (structural)	17	8 - 19
Aluminum	60	46 - 62
Copper	18	16 - 19
Zinc	13	12 - 14
Lead	11	10 - 12
Titanium	121	108-177
High-grade steel alloys	51	-
Recycled aluminum	10	9 - 12
Recycled copper	1.4	
Chemicals		
Ammonia (NH_3)	27	10 - 55
Soda ash (Na_2CO_3)	4	-
Sulfur (from pyrites)	3	-
Inorganic chemicals (average)	2.3	-
Miscellaneous		
Cement	2	-
Glass	6.2	-
Plastics (average)	2.5	-
Paper	6.0	3.8 - 7.5
Fired clay products (brick, terra-cotta)	6.0	3.5 - 14
Cotton fabric	10	9 - 13.5
Nylon fabric	22	19.5 - 27.5

Sources: Makhijani and Lichtenberg, "Conservation of Energy, Materials and Energy Use," in *Patterns of Energy Use in the United States,* Office of Science and Technology, Executive Office of the President (Washington, D.C., 1972). Raw data from *Census of Manufacturers,* 1967. R. N. Shreve, *Selected Process Industries* (New York: McGraw-Hill, 1950).

the manufacture of ammonia (and this list does not even include the electric arc process used by some countries) and concludes that "the design of an ammonia converter is still more or less of an art."[4] Clearly, to study the savings that might be derived by converting from one industrial process to another would be an incredibly complex job.

One can, however, make some generalizations. In the metals industries, recycling of scrap always consumes less energy than original manufacture. This applies to most other materials as well (although Hannon[5] has shown that such is not the case for glass). Thus, measures to encourage recycling will be repaid by energy savings, and will also conserve depletable resources.

The most effective means of persuading industry to reduce energy consumption is to increase the price of fuels relative to other goods and services. If the price of energy is increased substantially, processes using less energy will become the most profitable and will come into widespread use. If the cost accountants for a manufacturer find that overall costs can be minimized by adopting a less energy-consumptive process, that process will be adopted. Should prices be increased in this way, there will be additional incentive for recycling as well. Raising fuel prices is not likely to produce unemployment. As fuels become more expensive relative to labor costs, there will be less pressure to substitute energy and machinery for human labor.

How much energy can be saved by increased prices is not clear, but since manufacturing processes frequently differ by 100 percent in their energy requirements, we would think it possible to cut energy use in half in thirty years for the same output. A report by the Office of Emergency Preparedness estimates that industrial energy use can be decreased by about 6 percent by 1980, with little more than gentle pressure.

The figures in tables 9-1, 9-2, and 9-3 suggest means of achieving energy savings in transportation: transport people by bus and send freight by train, boat, and pipeline. While this is true as far as it goes, the problem is a little more complicated.

The private automobile is the largest single energy user in our society. In addition, large amounts of energy are used annually to construct highways. The energy cost of a mile of federally aided highway was about twice that of a mile of railroad track in 1970, and this energy cost has doubled since 1960.

Plans for more efficient engines are discussed, but prospects for a large improvement are slight. This is very strange, for we know very well how to cut the energy use of the automobile: reduce its weight and engine size. Small autos with small engines have long been available, but only recently have they become popular. Any serious effort to reduce energy consumption should begin by replacing large autos with small ones. It would take about a decade to replace most of the present stock. The energy saved would amount to about 7 percent of the United States total.

Since the average occupancy is less than two persons per auto, small cars would not necessarily cause a change in lifestyle. Increased fuel prices will undoubtedly provide pressure for the manufacture of smaller cars, but the industry will probably resist. Perhaps if all autos with engines larger than 150 horsepower were required to be licensed as trucks, we would have a further incentive to buy small cars.

Other savings in transportation are possible. A shift of freight from truck to rail would help, as would efforts to increase the service and availability of urban mass transit. Improved train service could reduce energy use in intercity passenger transport. Estimates of the effectiveness of these measures vary from about 5 to 12 percent of total United States energy use in 1980.

In the foregoing sections we have tried to estimate the amount of energy savings that might be possible without major changes in society. Table 9-6 summarizes our estimates alongside those reported by the Office of Emergency Preparedness. These two estimates differ largely in the conception of what is possible without major social and economic changes. Other estimates from the literature range from very small to more than 50 percent.

Why, then, have these changes not been made? Although energy is cheap, it has never been free. Now that we are conscious of our fuel shortages, and prices are increasing, can we expect these changes to occur naturally? It seems we cannot, for despite urban power shortages in recent years, there has been no substantial checking of the rate of growth in energy consumption.

This chapter cannot end on the note of optimism implied by the estimates of savings possible. Energy use has been doubling every fifteen years. If all the suggested energy conservation measures were to be accomplished in the next few years (probably an impossible feat), all we will

Table 9-6.   Estimates of Energy Savings Possible* with Existing Social Patterns

Area of Saving	Our Estimate	OEP Report
Residential/commercial	-	5.3
Residential heating	8.6	-
Commercial heating	1.2	-
Residential hot water	2.0	-
Defense	1.3	-
Industry	18	6.0
Small Automobiles	6.9	-
Transportation mode shift	5.0	5.0
Totals	43.0%	16.3%

* = in percent of total use

have done is postpone our problems by a decade or two. We most certainly should conserve, but, unless changes are made in the rate of growth, conservation measures will not be of much help. If we make the effort to save energy, will our society and our government be likely to the use the time gained to change society in ways that will slow growth in fuel use? The economist John Maynard Keynes was not optimistic: "I think with dread of the readjustments of the habits and instincts of the ordinary man, bred into him for countless generations, which he may be asked to discard within a few decades."

References

1. B. M. Hannon, *System Energy and Recycling: A Study of the Beverage Industry* (Urbana, Ill.: University of Illinois, Center for Advanced Computation, 1971).
2. U.S., Congress, House of Representatives, Hearings before the Subcommittee on Science, Research, and Development, *Energy Research and Development*, Min. 9-BO, 1972.
3. U.S., Congress, Senate, *Conservation of Energy, A Report to the Senate Committee on Interior and Insular Affairs*. Serial number 92-18 (Washington D.C., 1972).
4. R. M. Stephenson, *Introduction to Chemical Process Industries* (New York: Rheinhold, 1966).
5. Hannon, *op. cit.*

10

CLIMATE MODIFICATION AND ENERGY

The play seems out for an almost infinite run,
Don't mind a little thing like the actors fighting.
The only thing I worry about is the sun.
We'll be all right if nothing goes wrong with the lighting.

Robert Frost

The phenomena of weather and climate are expressions of the flow and conversion of solar energy. Matter and energy are constantly exchanged between air, land, and water and carried to all parts of the globe by two great circulating systems, the atmosphere and oceans.

The atmosphere is an unconfined mixture of gases moving ceaselessly above the earth. We divide it into a series of layers, each of which is characterized by certain physical properties and related phenomena (figure 10-1). The boundaries between these layers are neither sharp nor invariable, but rather, they represent regions of transition. The layer nearest the ground is the troposphere, seat of most of the phenomena collectively called "the weather." Ordinarily, there is extensive vertical mixing of air in the troposphere, and temperature decreases fairly uniformly with height. Of the total mass of the atmosphere, about 90 percent is in this layer.

The transition between the troposphere and the stratosphere is marked by a low point in the vertical temperature gradient, after which the temperature begins to rise. There is very little water vapor in the stratosphere, although cirrus clouds sometimes form in the lower part of it.

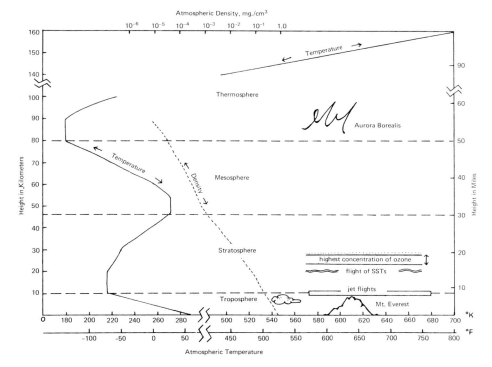

Figure 10-1. Structure of the atmosphere.

How the stratosphere influences weather and climate is not understood very well.

Above the stratosphere the temperature declines sharply with increasing height until it reaches a minimum about 50 miles above the earth. This region of decreasing temperature is called the mesosphere, and above it lies the thermosphere. The thermosphere is characterized by rapidly increasing temperatures which may reach 2500° F and by bands of ionized gases and free electrons (hence the other name for this region, the ionosphere). Major belts of ionized gas occurring at heights of about 60 miles and 150 miles are responsible for the aurora borealis — the "northern lights." Although changes in the upper atmosphere are related to long term climatic phenomena, we do not know much about the nature of the relationship.

The dry atmosphere to a height of about 50 miles is a mixture of nitrogen and oxygen with small amounts of other gases, as shown in table 10-1. In addition, the troposphere always contains water vapor in amounts varying from only traces in polar and desert regions to as much as 4 per-

Table 10-1.   Composition of the Atmosphere

Substance	Percent by Volume
Nitrogen	78.09
Oxygen	20.95
Argon	0.93
Carbon dioxide	0.034
Neon	0.0018
Helium	0.00052
Krypton	0.00011
Hydrogen	0.00005
Xenon	0.000008
Ozone	0.000001

Variable constituents: water, in solid, liquid, and gaseous state, averaging about 0.6 percent; solid particles, including dust, pollen, spores, and salts; carbon monoxide, nitrogen oxides, sulfur oxides, gaseous hydrocarbons, and other organic and inorganic molecules.

cent in the moist tropics. The troposphere also contains salts, primarily chlorides and sulfates, whipped into the air by the action of the wind on the sea. Pollen and spores and bacteria, dust from the surface of the earth and dust thrown out by volcanic eruptions, dust, chemicals, and smoke resulting from the activities of man, and hydrocarbons produced by plants are all part of the troposphere and all participate in the phenomena of the weather.

The energy that drives the winds and the water cycle and warms the surface of the earth is derived from the sun. Although solar radiation spans much of the electromagnetic spectrum, more than 90 percent of it is of wavelengths between 0.4 and 4.0 microns, corresponding to visible and infrared regions of the spectrum (figure 10-2). Solar energy in wavelengths less than about 0.2 microns is absorbed and converted to heat by oxygen in the thermosphere. Longer ultraviolet wavelengths, up to 0.3 microns, penetrate into the mesosphere and stratosphere before they are absorbed by ordinary molecular oxygen and ozone. By the time solar radiation reaches the lower stratosphere, about 3 percent of it, the shortest wavelengths, has been absorbed. Most of this has been converted directly to heat, but a part has caused ionization of gases or dissociation of molecules into atoms.

About 35 percent of the energy of wavelengths longer than 0.3 microns is reflected from the atmosphere and the surface of the earth. The fraction of reflected energy is called the albedo. The albedo of clouds ranges from a few percent for light cirrus clouds to more than 80 percent for dense cumulus clouds. On the average, it is about 50 to 60 percent. Since the average cloud cover of the earth is about 50 percent and the albedo of the clouds is 50 percent, we see that roughly 25 percent of the energy striking the atmosphere is reflected back into space by clouds. Any change in cloud cover would immediately affect the amount of energy received by earth. Similarly, a change in the distribution of clouds would affect the distribution of incoming solar energy.

Ten or twenty percent of incident solar energy is absorbed in the atmosphere, chiefly by water vapor and carbon dioxide. Radiation that finally reaches the surface of the earth is either reflected or absorbed. That which is absorbed is converted to heat and either stored, reradiated as long wavelength radiation, transferred back to the atmosphere by conduction or convection, or used to evaporate water. A very small part of it participates in photochemical reactions, including the most important photochemical reaction of all, photosynthesis.

Weather and cloud patterns for the entire western hemisphere are shown in this photo from the ATS III synchronous meteorological satellite. (Photo courtesy of NASA.)

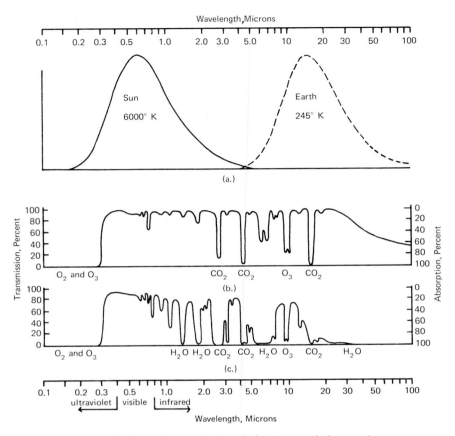

Figure 10-2. Radiative characteristics of the sun and the earth.

The radiative characteristics of a body depend on its temperature. The effective radiating surface of the earth is not its solid surface but a region of the troposphere whose temperature is about -18° F. Because the earth is relatively cool, its radiation is largely confined to the less energetic, longer wavelength region of the spectrum between 5 and 100 microns (figure 10-2). More than 90 percent of the energy radiated by the surface of the earth is of wavelengths that are absorbed by water vapor and carbon dioxide; much of it is absorbed or reflected back to earth by clouds. The transparency of the atmosphere to incoming visible light coupled with its opacity to outgoing infrared creates the greenhouse effect, which warms the surface of the earth. The effects of various constituents of the atmosphere on solar radiation are summarized in table 10-2. Anything we do to change the amounts of these constituents will affect earth's energy balance and consequently the climate.

The atmosphere is a great heat engine that runs on solar energy, moving the surplus of energy that falls in the equatorial zone to the poles, where there is a deficit. In general, warm air rises at the equator and moves toward the poles, where it is cooled, sinks, and moves back toward the equator again. The rotation of the earth makes things a bit more complicated, causing bands of prevailing winds: the polar easterlies, the westerlies of middle latitudes, and the trade winds near the equator. Further modifications in atmospheric circulation occur because of the irregular distribution of continents and oceans and of the location and topographic features of land masses.

Table 10-2.   Interaction of Atmospheric Components with Solar and Terrestrial Radiation

Substance or Material	Effect
Ozone (O_3)	Absorbs radiation of wavelengths shorter than 0.3 microns and converts it to heat. Reradiates this energy toward earth and into space, warming the stratosphere.
Water vapor	Absorbs radiation strongly in the range of 5 to 8 microns and greater than 18 microns. The average water vapor content of the atmosphere is about 1,500 trillion tons, which stores 9,000 trillion kilowatt hours of energy as latent heat. The average annual rainfall releases about 5000 times the world's annual energy use by man.
Clouds	Absorb between 2 and 30 percent of incoming radiation and reflect between 20 and 85 percent, depending on type of cloud and angle of radiation. Absorb long wavelength radiation from earth's surface and reradiate it back toward earth and out to space. Average cloud cover = 50 percent; average albedo of clouds = 50-60 percent.
Carbon dioxide	Absorbs radiation between 0.7 and 18 microns (most strongly between 12 and 16 microns). Reradiates it toward earth and space.
Oxygen, nitrogen and other atmospheric gases	Generally transparent to both solar and terrestrial radiation, although radiation in short wavelengths causes dissociation and ionization of molecules. They scatter radiation in a symmetrical way which depends on the wavelength of radiation and the size of the molecule; this is why the sky looks blue.
Oxides of nitrogen and sulfur, hydrocarbons, carbon monoxide	Negligible direct effect on heat balance, but some of these undergo photochemical reactions which involve ozone and/or which produce smog.
Particles (smoke, smog, dust, salts, pollen, spores, etc.)	Scatter and absorb solar and terrestrial radiation. Absorption occurs throughout the infrared. Scattering is asymmetric, depending on the size and shape of particles. Particles increase the albedo of the atmosphere in general, but may decrease the albedo of clouds; act as nuclei for cloud formation and condensation of rain and snow.

The oceans are another heat engine, running on the same principles as the atmosphere and interacting with it. Ocean currents are established by differences in temperature and density of water and are modified by the prevailing winds. Some 20 to 25 percent of the energy carried from equatorial regions to the poles is carried by ocean currents. In addition to the heat directly redistributed by the oceans, almost 20 percent of the energy carried poleward is in the form of latent heat in water vapor, most of which evaporated over the oceans. This heat is released in the atmosphere when rain or snow falls.

Thus climates of the earth are determined not only by amounts of energy received from the sun and radiated back from the earth, but by the way in which energy is distributed between the equator and the poles. When the distribution is fairly uniform, global climates are mild. When circulation between the equator and poles is restricted, an ice age may result. We do not have to blot out the sun to change the climate, or warm the earth by the sheer magnitude of our production of waste heat. We can do and perhaps are doing many more subtle things to change the winds, the rains, and the snows.

Modern man evolved during an ice age, a period of violent activity on earth — of mountain building, earthquakes, and volcanic eruptions. He has never known the earth as geologists believe it to have been throughout most of geologic time. In "normal" times, the earth is calm and mild. Polar seas are free of ice and polar climates are similar to those of middle latitudes today. There are vast expanses of continental lowlands covered by warm, shallow seas. The awesome mountains which seem so enduring are geologically quite young. The earth has repeatedly thrust up mountains like these, only to wear them down quickly and restore the low relief of the land.

There have been at least four ice ages, and maybe five or six, in the earth's history. All of them may have been associated with volcanic eruptions, violent contortions of the earth's crust, retreat of the seas, and expansion of deserts. Ice ages encompass alternating periods of growth and recession of glaciers, called glacial and interglacial stages. Scientists are not certain whether the period in which we now live is the end of the Pleistocene ice age or merely one of its interglacial stages. Conceivably, the answer to this question could hinge on the activities of man.

The concept of a stable climate becomes uncertain when we ask, stable over what period? Although names of places bring to mind general notions about their climates — London, Rome, Phoenix — climates everywhere show both cyclical and noncyclical changes. Cycles may be related to sunspots or the rotation of the sun, or to changes in the earth's orbit. Other changes cannot be explained satisfactorily. We may know that the major winds shifted their course without knowing why.

World and local climates have undergone many fluctuations since the last glacier receded eleven or twelve thousand years ago. These fluctuations are documented by the annual growth rings of trees, the age, depth, and character of annual deposits of silt in lakes, the succession of plant types in deposits of peat, the legends and histories of man, and, within recent times, instrumental measurements. These records show that about five thousand years ago the earth's climate was warm and humid, and it is possible that the Arctic Ocean was free of ice. Then began a period of gradual decrease of temperature and rainfall, but the trend was irregularly punctuated by long droughts. In Europe, where our knowledge is most complete, the main dry periods fell between 2200 and 1900 B.C., 1200 and 1000 B.C., and 700 and 500 B.C. The last drought was intense enough to cause extensive migrations of people away from the driest regions.

Then came a rapid change to a colder and wetter climate. Alpine passes which had been traveled for a dozen centuries were no longer passable. The change was abrupt and for many people catastrophic, but it did not last long. The Christian era was ushered in by conditions much like those of today.

Similar trends seem to have occurred simultaneously in Europe, western Asia, and North America. From about A.D. 180 to 350, it was rainy and rather cold. The fifth and seventh centuries were warm and dry; Alpine passes that had long been closed by ice and which are closed again today were heavily traveled. The ninth century bought heavier rains and cooler temperatures, but warm dry climates returned during the tenth and eleventh centuries. Greenland was settled in 984, but had to be abandoned to the ice about 1410. To this day, its climate is inhospitable to man.

The period from the seventeenth to the mid-nineteenth century was so severe that it has been called the Little Ice Age. During this time remnants of mountain glaciers crept far down into valleys and new glaciers formed. A warming trend began about 1850, melting the glaciers back to their sixteenth-century positions. Since 1940, however, average global temperatures have been decreasing. The change is greatest in the polar regions; the tropics have remained about the same.

Many explanations have been invoked to account for these many changes which, except perhaps for the most recent, were unrelated to the activities of man. None of the explanations is completely satisfactory. They include variations in the earth's orbit, disturbances in the crust of the earth, changes in composition of the atmosphere resulting from volcanic activity, changes in cloudiness whose causes can only be guessed at, and, for those who want to push the cause far from home, variations in the radiation from the sun. Whatever the causes, the only unchanging facet of climate is change. It has been a major, if not the major, force in evolution and in the rise and fall of human cultures. If conditions had remained ideal for the dinosaurs, there would probably be no men today to worry about man's influence on climatic change. There is no reason to suppose that our present climate will prevail. The question, then, is what is the natural trend, and how are we influencing it.

Homo erectus Pekinensis began to change the climate when he first carried fire to his cave. He changed the climate immediately about himself, the microclimate, from cold and damp to warm and dry. As man spread over much of the earth, as he changed from a hunter to an agriculturalist to an industrialist, his influence on local climate became greater and greater. There is no question about these effects; our bodies can sense them as we move from the central city to the suburbs to outlying rural areas, or from a forest to adjacent farmland. At some point, the cumulative impact of local changes will cause global changes. Here we are on much less firm theoretical and observational ground. We have great difficulty in determining the relative contributions of man and of forces beyond man's control. But the significance of our role behooves us to try.

In addition to putting excess heat into the atmosphere, man's use of energy can affect climate in a number of ways. Almost any change man makes in the landscape alters the albedo of the surface of the earth. It also tends to change the local balance between evaporation and precipitation, which may affect the distribution of clouds and the albedo of the atmosphere. By adding dust, smoke, carbon dioxide, water vapor, and a variety of chemicals to the atmosphere, man changes its reflective, absorptive, and radiative properties.

Some of the climatic repercussions of man's use of energy are related to the uneven distribution of his inputs into the environment. The rate of world energy use, which is essentially equivalent to world production of heat, is about 7 billion kilowatts. This is insignificant compared to the 90 trillion kilowatts of solar power that reach the earth's surface, or the latent heat added to the atmosphere at a rate of 40 trillion kilowatts through evaporation of water, or the energy of the winds flowing at a rate of about 400 billion kilowatts. Even a local thunderstorm releases energy equivalent to that of many hydrogen bombs. But the sun shines on half the earth at once, the winds blow everywhere, two-thirds of the earth is covered by water and much of the rest is covered by transpiring plants, and thunderstorms release their energy only occasionally at any one spot. Man's use of energy, in contrast, is heavily concentrated in urban and industrialized areas (table 10-3).

About 75 percent of the energy used by man is used in urban industrial areas covering only 0.1 percent of the earth's surface. All the energy used by man eventually finds its way into the atmosphere as heat. Although consideration of table 10-3 should put to rest any fears about heating up the whole earth with our waste heat, it is easily seen that we have heated up our cities. Table 10-4 shows the ways in which cities

Table 10-3.   Comparison of Man-made and Some Natural Flows of Energy

Energy Flow	Rate (watts/square foot)
Average solar radiation at earth's surface	approx. 15
Energy used to evaporate water, average rate	7.1
Average energy output of urban industrial areas	1.1
Average net photosynthetic production on land	0.012
Average heat flow from the interior of the earth	0.0049
World use of energy (1970), distributed evenly over the continents	0.0042
World use of energy (1970), distributed evenly over entire earth	0.0012

modify local energy balance and also the climatic effects of these modifications. Waste heat and pollution — the two items related directly to use of energy — have by far the greatest effect on local climate.

Our use of energy produces both sensible heat and latent heat. Latent heat is stored in water vapor, through evaporative cooling processes. It does not affect the temperature until it is released when the water vapor

Table 10-4.   Climatic Effects of Urbanization

Urbanization produces many changes that affect energy balance. The albedo of the surface is increased due to the reflectivity of buildings and pavement. In snowy areas, however, there may be a compensating decrease in albedo due to dirty snow and snow removal. Accompanying the change in albedo is a change in heat storage capacity of the surface. Buildings and other obstacles to wind flow create increased atmospheric turbulence near the surface. There is decreased evaporation and transpiration of water and increased runoff of precipitation. Burning of fuel produces air pollution, both gaseous and particulate, and waste heat. These changes are related to the following observed climatic effects:

Typical differences of urban climates from surrounding rural areas

Cloudiness
    5-10% more cloud cover
    100% more winter fog
    30% more summer fog

Precipitation (effect may be more pronounced downwind from urban area, at times)
    5-10% more (total)
    5% more snow

Relative humidity
    2% less (winter)
    8% less (summer)

Radiation striking surface
    15-20% less
    30% less ultraviolet in winter
    5% less ultraviolet in summer

Temperature
    0.5°-1.0° C higher annual mean
    1°-2° C higher winter minimum

Winds
    20-30% lower mean annual wind speed
    10-20% decrease in extreme gusts
    5-20% increase in calms

condenses. Sensible heat creates a dome of warm air that hovers low over the city. Released near the surface it increases the temperature gradient near the surface, consequently increasing the rate and extent of vertical movement and mixing of air. Latent heat, in contrast, rises higher and travels further, and may show its effect many days later, far from its source.

The dome of warm air is also a dome of polluted air. When the wind blows, a plume of warm, polluted air is carried downwind from the city. On this scale, changes induced locally begin to affect the surrounding countryside. However, incomplete understanding of how regional climatic patterns are determined prevents scientists from estimating the impact of cities on the climate of their surroundings.

At some point, as cities proliferate and expand and merge with one another, their effects will be widely felt. A number of large areas, notably northwestern Germany and southern Belgium, the northeastern United States, and southern California, are putting waste heat into the atmosphere at a rate between 1 and 10 percent of the rate at which they receive energy from the sun. Heat output per unit area is not expected to grow very much, but the size and number of the urban industrial areas will.

Until recently, the amount of waste heat produced was much greater in the winter than in the summer, as a result of space heating. In cities like Moscow and Fairbanks, which have very cold winters, man may approach or surpass the sun's rate of input of energy to the local environment. Further south, however, increasing use of air conditioners has significantly increased the production of heat during warm months. In this sense, the more air conditioners we have, the more we need. Moreover, the trend toward heating the outside in winter may be slowing down, with new emphasis on effective insulation of buildings and increased use of heat pumps which can warm the interior of a building with heat removed from the outside.

The unsuspected energy subsidies to agriculture and the myth of one farmer feeding fifty people were discussed in chapter 4. Agriculture, whether modern or primitive, also affects the energy balance of the earth. While the magnitude of the effect is smaller per unit area than the effect of urbanization, vastly greater areas are involved.

Conversion of forest or grassland to agricultural land influences the local energy balance in three ways. It changes the albedo of the surface, it changes the relative proportions of sensible and latent heat in the total heat rejected from the surface, and it often increases the amount of dust in

A satellite photo of inadvertent weather modification. Smoke and other emissions can be seen streaming northeast from industrial complexes at the south end of Lake Michigan. As they reach the clouds, the emissions produce heavier cloud ridges. (ERTS photo from Walter Lyons.)

the atmosphere. The conversion of temperate forests to agricultural land may have gone about as far as it will go, but large-scale cutting of tropical forests in underway. When tropical forests are cut, moisture can no longer be retained in the soil. Flash floods cause severe erosion. During dry seasons, clouds of dust may be released into the atmosphere. The ratio of sensible to latent heat increases owing to greatly reduced rates of transpiration by plants. Incoming solar energy is diverted toward warming the surface rather than toward evaporating water. This means that there is more energy immediately available to drive atmospheric motions, the thermal gradient in the lower atmosphere becomes steeper, and vertical air movement increases. Some meteorologists believe that altering the ratio of sensible heat to latent heat over large areas of the tropics may affect the flow of the tropical easterly winds and in this way modify the general circulation of the atmosphere.

About 1.5 percent of the total area of the continents is estimated to be under irrigation. In desert regions, irrigation has a profound effect on the heat balance because of the increased rate of evaporation and excess of evaporation over precipitation. Here the effect is opposite to that of cutting tropical forests. Increased evaporation causes a decrease in the amount of energy immediately available for atmospheric processes. The thermal gradient in the lower atmosphere decreases. Latent heat becomes available days or even weeks later, when the water vapor condenses and precipitates. The energy is released far from where the evaporation took place, and perhaps it is released high in the sky. Evaporation of water is a process that tends to export solar energy to places remote from where it struck the earth. Worldwide, the decrease in transpiration due to the cutting of forests is thought to be of about the same magnitude as the increase of evaporation due to new irrigation, but the two effects cannot compensate for each other because they occur in different parts of the world.

Artificial lakes are frequently formed in conjunction with the generation of electricity, either at a hydroelectric site or as cooling ponds to dissipate waste heat from the condensers in a steam electric plant. The major effect of an artificial lake may be similar to that of irrigation. It leads to increased evaporation without a corresponding increase in precipitation, and increases the latent heat released to the atmosphere (see the discussion of thermal pollution in chapter 8). Manmade lakes and reservoirs are scattered over the earth, however, and are widely separated from each other. Their total area is only about one-tenth the irrigated area of the world. Thus, their climatic effects are expected to remain small and local.

Determining local climatic effects of human activity is relatively easy. Changes in cloudiness, temperature, rates of evaporation, and other variables can be measured and related to burning of fuel, cutting of forests, and other changes made by man. For comparison, there are data from surrounding areas not directly modified by man (or modified less drastically, or in a different way) and from the affected area before man changed it. The situation for global effects is quite different. The sequence of cause and effect is seldom obvious, and at present it is impossible, in many cases, to determine the relative magnitude of manmade and natural effects. The problem is compounded by our uncertainty about what the natural trend in world climate is.

We do, however, understand the basic dynamics of the atmosphere and we know how individual factors modify the amount, distribution, and flow of energy. We are beginning to understand the many ways in which these factors interact. Climatic studies have frequently been hampered by lack of reliable data, especially long-term records. Data are now accumulating from weather satellites and other sources faster than they can be interpreted. With these data and existing theory, it is possible to formulate hypotheses about climatic change and to test them by computation. When climatic features are represented by mathematical equations arranged for solution by a computer (as, for example, the rate of increase in atmospheric carbon dioxide and the relationship between concentration of carbon dioxide and surface temperature), we have a mathematical simulation or model of the atmosphere. These models, imperfect as they are, are the best means we possess to learn about the tangle of relationships which govern climate and the behavior of the atmosphere. As one meteorologist put it, ". . . we can eventually say what features or combinations of features could have produced the changes. . . . As to what features did produce [them] we shall still have the privilege of arguing." This may be said for predictions as well as for hindsight.

The best current models, requiring hours of time on the largest computers, are still unable to take all relevant factors into account. They accommodate changes in the water content of the atmosphere, but they cannot say whether this water will result in increased cloudiness, increased precipitation, or both. They do not represent adequately the effects of variable cloud cover and interactions between atmosphere and ocean. Different models may lead to conflicting results and paradoxes — a warming trend predicted by one and a cooling trend by another. Of course these problems did not originate with the use of computers. They reflect the limits to our understanding of natural processes and arise from assumptions we make about the system; they cannot be blamed on the computer or the technique of modeling. It has been suggested, for example, that an ice age could be started by an increase in solar radiation, which would lead to increased evaporation of water, increased cloudiness, increased rain and snow, and decreased temperatures. Most people, however, do not believe that a hotter sun would cause a colder earth.

Aware of the problems and promises of climatic models, we will look at some of the observed and predicted changes in variables affecting the climate and their possible outcomes.

There are about 2.3 trillion tons of carbon dioxide in the atmosphere. Each year we release an increasing amount through combustion of fossil fuels. In 1950 it was 5.8 billion tons, by 1965 it had doubled to 12.2 billion tons, and for 1980 we estimate that it will have doubled again to 23.6 billion tons.

The first systematic measurements of carbon dioxide concentration were begun in 1958. They reveal a number of interesting facts. In the northern hemisphere there is an annual oscillation in carbon dioxide concentration which is related to the cycle of photosynthetic productivity. However, the mean annual concentration seems to be growing everywhere at a rate of about 0.2 percent per year. The highest levels are reported from Point Barrow, Alaska, and the lowest from Antarctica. In general, values in the northern hemisphere are higher than those in the southern hemisphere. This might be expected, since most fossil fuel is burned in the north. The difference suggests a time lag of about eighteen months in the mixing of air between the arctic and the antarctic.

Almost half of the carbon dioxide added to the atmosphere by man remains there. The remaining half must be taken up by the oceans or the biosphere — there is nowhere else for it to go. We do not know how much of the excess may be assimilated by plants. In any case, the biosphere is only a temporary sink, because increased photosynthetic production leads to increased breakdown of organic material by herbivores and decomposers, which, in turn, consume oxygen and release carbon dioxide. It is generally assumed, for the purpose of designing climatic models, that about half of each increment of carbon dioxide added to the atmosphere will continue to be removed; but most natural processes do not remain so nicely linear. We assume that this one will remain linear only because we do not know how it is likely to change. In the short term, the assumption is accurate enough.

Simplified models of global climate predict that a doubling of carbon dioxide concentration would increase the average surface temperature by about 3.6° F, while the stratosphere becomes cooler. The increase in carbon dioxide that is predicted for the year 2000 would cause a warming of about 0.9° F. As temperatures increase absolute humidity will also probably increase, and infrared absorption by the additional water vapor will contribute to the warming effect; this model takes these facts into account. But as usual, the entire picture is not this simple. We do not know, for example, if the increased humidity will produce more clouds. The temperature increase will probably be greater toward the poles than in the tropics, for as the snow cover recedes the albedo of the surface will decrease and more solar energy will be absorbed. Decreased temperature differential between the equator and the poles has implications for global patterns of atmospheric circulation. It is possible that the Arctic Ocean would become and remain essentially free of ice. If this should happen, patterns of oceanic circulation would also be profoundly changed. On the other hand, if the increased content of water vapor in the atmosphere leads to an increase in cloud cover of only 0.6 percent, the cooling effect of the clouds could negate the warming effect due to carbon dioxide.

We could halt the atmospheric increase of carbon dioxide by turning to sources of energy other than fossil fuels. But no matter what progress is made in the technology of nuclear and solar energy, man will probably be burning large amounts of fuel for at least several centuries to come. By the time we realize that carbon dioxide is causing a problem, it may be too late to do anything about it. Once a trend toward climatic change has started, halting the increase of carbon dioxide is unlikely to stop it because of the way in which the effects of decreasing snow and ice cover and altered patterns of circulation reinforce each other.

Apart from sudden increases in the particulate content of the atmosphere following major volcanic eruptions, there appears to be a global trend toward increase of particles. This conclusion is based largely on measurements of solar radiation reaching earth's surface and on the assumption that decreases in solar radiation are due to increased scattering and reflection by atmospheric particles. In cities, there has been a well documented 15 to 20 percent decrease in solar radiation during the last 50 years. Observations are neither consistent enough nor of long enough standing in nonpolluted areas to come to firm conclusions, but evidence suggests that in the northern hemisphere solar radiation has been reduced by about 5 percent even in areas of "clean air." Measurements over the North Atlantic suggest a doubling in number of particles between 0.02 and 0.2 microns in diameter. Because these data were collected in an area undisturbed by human activity, they may represent an overall trend in the northern hemisphere. There are few reliable observations for the southern hemisphere, but the concentration seems to be smaller there, as would be expected because the major sources of manmade particles are concentrated in the north.

Models of climatic effects of particles must take into account the particles' horizontal and vertical distribution, their size, and their optical properties. Because information of this sort is inadequate, predictions from climatic models are tenuous at best. The most that can be said with confidence is that an increase in particles will increase the albedo of the atmosphere and will probably tend to cool the surface of the earth. How much our production of particulate matter will influence the climate remains a vital but unanswered question. It is comforting to note, however, that with a shift to nuclear or solar energy and with stringent pollution controls and attention to agricultural and mining practices, almost all manmade atmospheric particles could be eliminated. Because the lifetime of particles in the atmosphere is relatively short (unlike the lifetime of excess carbon dioxide), the problems of air pollution will not linger to haunt generations of people who may no longer be polluting the air.

Potential changes in composition of the upper atmosphere are related to the effects of jets flying in the upper troposphere and lower stratosphere, supersonic transport aircraft (SSTs) flying in the stratosphere, and rockets, which penetrate all layers of the atmosphere.

By 1990, jet aircraft flights may increase sixfold and more, if SSTs do not begin operation as expected. There is no doubt that condensation of water vapor emitted from these aircraft (causing the infamous "contrails") will increase the cloudiness of the upper troposphere. The effect will be most pronounced in heavily-traveled air corridors near major jet airports. The primary effect of increased cloudiness will be to increase the albedo of the atmosphere, although it is difficult to estimate how large the effect will be, or how it will affect the surface of the earth. It is also possible that ice crystals from the upper troposphere will fall into clouds below, seeding them and causing precipitation to occur sooner that it otherwise would.

Concern about the effect of SSTs involves their emissions of water vapor, carbon dioxide, and particles. The best experience we have to go on, related to modification of the stratosphere, is the known effects of major volcanic eruptions. Injection of volcanic dust into the stratosphere causes a rise in stratospheric temperature of as much as 11° to 13° F. Increased volcanic activity also appears to be associated with decreased surface temperatures. A strict analogy cannot be made between volcanic emissions and the emissions of SSTs, however, because SSTs will operate within a narrow range of altitude and primarily in a few heavily traveled air corridors in the northern hemisphere. Volcanic debris is more widely distributed by latitude and longitude and through all vertical layers of the atmosphere.

The greenhouse effect of stratospheric water vapor and carbon dioxide is expected to be very small, and insignificant relative to the effect due to burning of fossil fuel. Because the stratosphere is very dry, it is expected that additional water vapor will cause increased cloudiness only in the coldest regions. The idea that the ozone layer will be destroyed through increased photochemical reactions involving ozone, water vapor, and other emissions has been carefully studied and discredited, because the aircraft will fly below the most concentrated ozone layer and because computer models indicate that water vapor has the greatest effect on concentration of ozone at altitudes higher than where most of the ozone is found. In conclusion, those who have studied the problem most thoroughly say they do not think stratospheric flights of SSTs will cause any serious problems, but they are extremely reluctant to try the experiment and find out.

The higher we go in the sky the shakier our theoretical and observational bases for prediction become. Because there is so little matter in the upper regions of the atmosphere, each rocket and each explosion of a nuclear bomb in the atmosphere can cause a significant change in the composition of the mesosphere and thermosphere. We know that we are changing the chemistry and perhaps the circulation of the very high atmosphere, but we do not know what the effects of this change may be.

When we think of modifying the climate, we often think in melodramatic terms of another ice age, with all living things fleeing in advance of the glaciers. Or we think of melting the polar ice and flooding all our coastal cities. These would be but the final results of our activities. Many other changes would come before we froze or drowned. A slight but persistent change in temperature or humidity can tip the balance in favor of organisms pathogenic to plants, animals, or man, or of vectors (such as the mosquito) that spread those organisms. The result could be severe epidemics of diseases and pests that formerly were held in check by climatic conditions. Major epidemics of crops and forest trees have often been associated with just such changes. Or the major winds could shift their course, as there is some evidence that they are doing. They could fail to bring rain at the usual season or bring disastrous rains at the time of harvest, and the resulting crop losses would bring starvation to additional millions of people. We have always been slow to adjust agricultural practices to changes in climate.

Nor are climatic changes necessarily for the worse. If we really thought they were, we would be foolish to engage in research on practical applications of weather and climate modification (although it would be foolish on other grounds to undertake massive experiments of this nature*). The explanation for the fears of many of us is related to the feeling that danger lies in the unknown. And after the unknown becomes known, it may be too late to do anything about it.

---

* It would seem at first to be an unmitigated good, for example, to break up or detour the hurricanes that ravage our southern coastal states. On second look, it turns out that in so doing we would modify patterns of rainfall along the entire Atlantic seaboard, with results not quite so delightful as the successful conquest of tropical storms.

11

POLICY FOR ENERGY

OR ENERGY FOR POLICY?

*The late twentieth century stands in direct contrast to the expec-
tations of the late eighteenth century. Individuals depend in every
way on huge organizations, be they corporations or government
agencies. There is concentration of means, of decision making, of
power upon the future, of responsibility.*

Bertrand de Jouvenel

You boil it in sawdust; you salt it in glue;
  You condense it with locusts and tape;
Still keeping one principal object in view,
  To preserve its symmetrical shape.

Lewis Carroll

There was a time, not so long ago, when the five-foot shelf of Harvard
Classics was a common possession of those who aspired to be persons of
culture. Nowadays, perhaps reflecting a change in social values, a new
five-foot shelf adorns the walls of many offices. Its concerns are political
and economic, rather than literary and philosophical. The authors are corporate now—the National Industrial Conference Board, the National
Academy of Sciences, Shell Oil Company, the Sierra Club, a half-dozen
Congressional committees, the National Petroleum Council, American
Mining Congress, the National Wildlife Federation, and a host of three
and four letter non-words: NASA, NSF, API, USGS, OST, SRI, RAND,

and FOE—and the subject is energy policy studies. This new five-foot shelf, like its predecessor, goes largely unread, even when arguments are most heated. It serves mainly as a security blanket in the face of problems, and as a last solace that someone is seeking solutions.

There is little consistency among these studies. Some comments about energy policy are meant to show that our problems are not only serious, but insoluble. Other studies are meant to reassure us. Some are the comments of tough-minded experts who thrive on crises, and tell us that our hopes and dreams must be subordinated to this crisis. Some comments come from fuzzy utopians, who see, in energy shortages, a chance—at last—to perfect mankind, and solve the shortages too. Some comments are self-serving, some are altruistic, and a good many are just plain wrong. Appeals are made to free enterprise, social darwinism, self-restraint, more government regulation, less government regulation, and the American way of life. The cause of our present plight is variously given as big business, consumers, government, Republicans, Democrats, environmentalists, economists, scientists, engineers, Arabs, communists, capitalists, and the structure of society itself. The outcome (after a period of travail, of course) is variously predicted as a collapse of society, a new dawn of

Against the backdrop of an electric power plant's smokestacks, a crane adds to the growth of our modern society. In the halls of government, policies to reconcile the increasing pressures of growth and energy shortages are sought. (Photo courtesy of Marine Studies Center, University of Wisconsin.)

technological utopia (with energy too cheap to meter), final disruption of the earth's ecosystem, decline in the standard of living to some kind of minimal Malthusian misery, or war.

It's time to back off! It is worth a try to step back, to inquire just where we are headed anyhow, to identify the dimensions of the problem for which a policy is sought. The title of this chapter poses the question from which we may start. There is really only one answer. Society must have a larger policy first—a goal if you prefer. Then, and only then, it is possible to inquire whether this policy is in any way limited by available energy. If it is, we can find ways of supplying the necessary energy, or, if necessary, change the goal to fit the limitations.

The issue of "energy policy" has been introduced in this way to show in sharp relief the hidden agendas for society that lurk in the slogans of the various protagonists. Sometimes these slogans are painfully obvious. For example: one of the major oil companies tells us at every turn that "a country that runs on oil cannot afford to run short." The hidden goal here is nothing less than rigid adherence to the present state of affairs, denying historical differences and future possibilities (and incidentally illustrating an uncommonly low opinion of the public served). Why not "A country that runs on oil had better seek another energy source" or "A country that runs on oil ought to consider walking more?"

There are only two inputs to industrialized societies: energy and materials. Many materials can be substituted for one another. We can—and do—substitute aluminum for steel or glass. Similarly, we can substitute one fuel for another when energy is needed. But energy cannot be substituted for materials nor materials for energy. Thus, a continuous supply of energy is absolutely essential for an industrialized society. The oil embargo by the Arab nations in 1973 showed with dramatic suddenness how the control of energy by a few can apply enormous pressure to world politics. In an industrialized society *any* goal requires energy for its implementation. To find out how much is needed and the conditions under which it can and should be supplied is everybody's business.

Below are what we consider to be the minimum societal goals that must be served by any energy policy. Because the purpose here is to find out about energy limitations, if any, and about needed policy and institutional changes, we will exclude matters which do not place demands on energy resources, even though we may feel strongly about society's need to change. These minimum goals are: (1) adequate food, clothing, and shelter for all; (2) opportunity to share in whatever surplus goods and services remain after basic needs have been satisfied; (3) ample employment or other meaningful activity for each person; (4) freedom to practice any lifestyle, subject only to the constraints of the above goals.

Because the United States is but one of many nations sharing the resources of this planet, we must set our goals in the larger context of the world community. The constraints placed on our goals by analogous goals for the world would include:

(1) Anticipation of the material and energy needs of other countries so that U.S. actions do not deny them the opportunity to fulfill their own goals.

(2) Active participation in programs to lessen world tensions, to assist poor nations with solution of their problems, and to bring about equitable distribution of the world's resources.

(3) Active participation in programs to diminish threats to world ecosystems.

(4) Choice of technologies and policies that offer the opportunity to less developed nations to diminish, rather than increase, their dependence on the industrialized nations.

The ultimate aim of the above goals and constraints is nothing less than a life of dignity, meaning, and satisfaction for each person in a world at peace. Though the importance of this aim to policy makers is easily seen, its translation into actual policy formulations is much more difficult. It is hard to imagine many, liberal or conservative, who would quarrel with the basic objectives. They must, however, state their own basic goals explicitly, and reconcile them with the above constraints, before asking that their policy suggestions be adopted.

There appears to be an increasing concentration of control over U.S. energy resources. Let us look at the evidence for this comment.

The early days of the oil industry were characterized by free-wheeling competition which was often ruinous to the losers. The final result was the emergence of the large oil corporations. It was against these corporations that the early antitrust actions were undertaken. Competition again became vigorous after the collapse of the stock market in 1929. But during the prolonged depression of the 1930s, the price of oil dropped so low that many small companies went broke. Since then, through merger and purchase, a few corporations have gained control of our oil and gas resources. In 1969, the seven largest companies controlled more than 70 percent of the world petroleum market, and the twenty-five largest companies controlled almost 85 percent of the world market. Much of the remaining world market is controlled by the nationalized oil companies of other countries.

Industrial groupings of this sort do not respond to the free market mechanisms of classical economics. Each company's share of the market is sufficiently large that its own actions affect the overall price structure as well as its own competitive position. Such oligopolistic or monopolistic industrial groupings are not uncommon in our present society (steel is another example). When such industries have undifferentiated products, as do oil and steel, they inevitably engage in price fixing, whether by actual collusion, by following each other's lead in prices, or through some sort of official or unofficial regulatory mechanism.

Each of the twenty-five largest companies produces natural gas as well as oil. The price of natural gas is fixed directly by the Federal Power Commission. Until 1971 oil prices were determined by the production control sanctioned by the Interstate Oil Compact Commission. Although the state regulatory bodies which make up the Interstate Compact Commission do not fix prices directly, the effect is the same.

What happens is this: Production quotas are derived from estimates of the major oil companies of what they will market from each state that produces oil. These estimates are combined with estimates of demand that are prepared monthly by the U.S. Bureau of Mines. But estimating demand depends upon knowing the price, because the amount bought, especially by large consumers, varies with the price. The price used in all these estimates is always the same as the present market price or one slightly higher. Now if the demand and supply functions are known—and they are by the professionals who make such estimates—then the assumed price will always turn out to be the actual average price. Thus, the Interstate Compact Commission mechanism, which started out as a solution to the overproduction problems of the early 1930s, has ended in being a mechanism for price fixing as well as production management. Ended is the correct word, because in the spring of 1972, for the first time since production controls were begun, all wells were put on full production. We could no longer produce as much as was wanted. Notably, when this happened, domestic oil prices were fifty cents per barrel higher than the world market price, showing that the industry had not suffered unduly from the regulatory mechanism.

Most large oil companies are vertically integrated. This means that they control refineries and wholesale and retail outlets as well as production and exploration. In each of the last twenty-five years the eight largest companies controlled between 55 and 60 percent of U.S. refinery capacity. The ten largest oil companies operate or control retail dealerships. Eight of the ten operate in forty-three or more states. And make no mistake, retail sales of gas and oil are big business. Service station sales account for 7.5 to 8.0 percent of all retail sales.

In 1972 and 1973 several small independent refineries were operating far below capacity or were forced to close down owing to the lack of crude oil. Congressional hearings have been punctuated by persistent claims that independent retailers have also been forced to close or reduce business because of supply shortages. The best that can be said of these events is that the major integrated companies must be expected to serve their own franchises first, and the independent refiners and dealers last. At worst these events constitute monopoly activities in restraint of trade. In any case, there is little to be seen of free markets or competition.

Table 11-1.   **Production, Planned Production, Leasing, or Other Entry into Other Energy Sources by Largest Petroleum Companies**

Energy Source	Number of Companies with Active or Planned Production	Companies with Leases or Other Commitments	Total
Oil Shale	3	14	17
Tar Sands	3	13	16
Coal	7	9	16
Uranium	6	18	24

Source: L. C. Rogers, "Oil-finding Talent Pours into Broad Minerals Drive," *Oil & Gas Journal,* February 24, 1969, p. 37.

In the early 1960s diversification came to the oil industry. This meant entry into other forms of energy supply, or the acquisition of companies supplying other types of fuel. By 1971, major oil companies owned more than 30 percent of U.S. coal reserves and accounted for 20 percent of production; they owned more than half the uranium reserves and 25 percent of the uranium milling capacity; and they operated or were planning 93 percent of the nuclear fuel reprocessing plants. The continuing pursuit of other energy resources by oil companies may be seen from table 11-1. Remember, too, that all of these companies produce natural gas as well.

The oil companies—perhaps better called energy companies—are aware that we will turn to uranium-based energy sources in the future and accordingly, they have moved into this field (table 11-2). Many companies are seeking to enter, or have already entered, several phases of the supply and processing of uranium fuels.

At present, the heavily regulated electric utilities are faced with little interfuel competition to hold prices within reason. It is of no advantage to the energy companies to compete with themselves when they supply all fuels. Table 11-2 shows how little this situation will change if and when nuclear fuels begin to supply a substantial portion of our energy needs. Before Congress the chairman of the board of directors of the Tennessee Valley Authority told of one oil company with coal holdings that reported it had no intention of supplying TVA with coal unless the agency would pay a price that would yield the same return on coal as on oil. The oil companies have always maintained that they were entitled to a higher return than other industries because of the discovery risks they face. Whatever the merits of that argument, there is no such risk in coal. Yet this statement was made in 1970, prior to the fuel shortages that have strengthened the hand of the energy companies.[1] By the winter of 1972-73, cities, school districts, and other public agencies were widely reporting failure to obtain competitive bids for fuel supply—or any bids at all. What this meant was that the energy companies were prepared to take advantage of fuel shortages to deny bulk purchase economies to large-volume public buyers.

One oil company holds leases on more than three-quarters of all the

Table 11-2.   Present or Future Capability of Eight Large Oil Companies in the Nuclear Industry

Company	Exploration or Reserve Holdings	Uranium Mining & Milling	UF_6 Conversion	Fuel Preparation or Fabrication	Fuel Reprocessing	Reactors
Standard Oil	X	X		X	X	X
Gulf	X	X		X	X	
Atlantic Richfield	X		X	X	X	
Continental	X	X				
Getty	X	X		X	X	
Standard of Ohio	X	X	X			
Kerr-McGee	X	X		X	X	
Sun	X	X				

Source: B.C. Netschert, "The Energy Company: A Monopoly Trend in the Energy Markets," *Science and Public Affairs* 27 (1971): 13-17.

known geothermal prospects for energy supply. Thus, the oil companies are gaining both vertical and horizontal control of energy sources, processing, and marketing. Bruce Netschert, an energy economist for many years, regards the situation this way: "The development of the energy company thus presages fuel markets dominated at both the supply and the consumer levels by firms of immense size and monopoly power. . . . Far from being confronted with the limitations on market power—indeed, even dissolution and divestiture — that a sound competition — preserving antitrust policy might be expected to decree, the energy companies have been able to grow apace, free of the regulation that governs their utility competitors. and unimpeded by the strictures of antitrust policy to which many of their less powerful competitors have frequently been subjected."[2]

Before fuels became scarce, the concentration of power in a few large corporations attracted little attention from the public. The aims of the energy companies were mostly realized and the public had a ready supply of relatively cheap energy.

The potential danger posed by the concentrated control of energy can only be compared with the threat posed by a monopoly of material resources. Imagine a situation in which a few large corporations had control over metals, food, timber, synthetic and natural fibers, autos, trucks, trains, bicycles, cement, sand, gravel, chemicals—in short, over all materials and manufactured goods. That is nearly the case, for there is no material that can be substituted for energy. Thus, it is crucial to inquire about the priorities of the energy companies.

In 1972, the American Petroleum Institute (API), primary trade association for the largest energy companies, issued a statement of energy policy which indicates these priorities. It begins with a statement of principles which are purported to be "essential for the protection of long range national interests." The statement calls for: (1) "Maintenance of a substantial domestic energy base," which is said to be vital for national security and for reduction of our balance of trade deficits; (2) "Increased development of domestic energy resources . . . and increased efficiency of utilization"; (3) "Policies . . . to insure continuance of the vital energy flow . . . should be developed by government in consultation with industry." It is asserted at this point that "the market forces of a competitive economy will provide the surest and most economical means to develop and allocate the needed energy supplies;" and (4) Balancing the "national goals of protecting the environment and developing secure energy resources required to meet the nation's essential needs at reasonable cost."[3]

Most of these principles are sufficiently general that everyone could accept them, but there are several misleading implications. First, there is no recognition that conventional fuels simply cannot meet long-range U.S. needs. Second, the appeal to a "competitive economy" and "market forces" would suggest that the American Petroleum Institute is unaware that control of energy resources is concentrated and that the usual competitive forces of a free market do not exist in the energy industry. Third, the idea of balancing environmental needs and energy needs could signal a search for a careful course between conflicting goals, but the recommendations following the statement demonstrate an unmistakable thrust toward an energy first, environment second policy. The API is concerned with policy for energy, and not with energy for policy. Perhaps, the real clue is their insistence that when making policy government should consult with industry, and not with labor, citizens' groups or other countries.

The Institute's policy recommendations offer a more specific indication of the energy companies' priorities. These recommendations fall into six categories, which are summarized below.

(1) Recommendations on exploration and development begin with the flat statement that "government should not engage in oil and gas exploration and development," and that research and development should be a private responsibility. It is stated that "federal lands (i.e. public lands) *must* be made more readily available for exploration and development," and that "U.S. jurisdiction over the seabed . . . to the outer edge of the continental shelf should be retained," and "the government *must* give its full support to the competitive overseas efforts of U.S. companies." The juxtaposition of these statements should make clear one set of industry priorities: the government is to help when asked, but otherwise to keep its hands off. Whether the government should engage in exploration is not clear, but if the government does not explore, how are we to know what the nation's supply of fuel resources really is?

Through the most powerful lobby in Washington, the energy companies have been able to exempt themselves from supplying any information on the results of their explorations. The Public Information Act of 1968 (the Moss act) specifically exempts oil company data from being made public, and, in fact, prohibits the government from releasing whatever information it may receive from the energy companies on oil and gas discoveries. Thus the stricture against exploration asks that the government accept only the information which the energy companies choose to supply, and seek nothing more.

The industry demands that public lands be made available for exploration. Public lands belong to everyone, not to an abstraction called the government. Any prudent person ought to know what these lands contain, and what uses might be made of them before selling off mineral exploitation rights. The government, as a representative of the owners, has an obligation to know what it is selling.

The recommendations for "full support" of private overseas efforts, and for the assertion of territorial rights over the continental shelf, are direct requests that U.S. foreign policy be made in accordance with the desires of the energy companies. Possibly the energy companies look with envy on the close ties between Japanese industry and the Japanese government. These ties have enabled them to make deals for industrial assistance in exchange for guaranteed oil supplies. But are U.S. oil companies ready to accept the government control that goes with such a policy? Are they ready for employment control and profit control? All too often in the past, full support of U.S. corporations abroad has forced our government to support unpopular foreign governments, in direct contravention of our own national principles. If we have not learned by now the futility of meddling in the internal affairs of other nations, we are doomed to repeat the national trauma of the Vietnam war over and over again.

Freedom of the seas has been the best example of international trust in a distrustful world. Yet even that threatens to slip away from us. First in Malta and more recently in Iceland, Peru, Canada, and elsewhere, nations have laid claim to continental shelves and beyond. In the Middle East, such an extension of sovereignty into the Gulf of Aqaba led to war in 1967. Perhaps seabed resources should be used as the basis for new international agreements with the aim of world peace. In any case, the United States, richest nation in the world, should not make policy on the narrow ground of short-term energy needs, and close out our fishing, shipping, and mining interests in the bargain. Do we really want the continental shelves of the rest of the world closed to us in reprisal?

(2) A second set of recommendations relates to taxation and regulation. Four of these recommendations have a single thrust: reduce regulation and increase tax credits and other "incentives."

The energy companies seek an end to regulation of natural gas prices by the Federal Power Commission. They believe, and many outside the industry agree, that natural gas prices have been held to an unrealistically low level. Prices probably should be raised in order to shift some of the largest users away from natural gas as a fuel source. But the energy companies want government regulation of prices to cease altogether. API states that gas "should be allowed to compete freely in the marketplace"—which might be a good idea if there were a free market for natural gas. But can the homeowner with a gas furnace make any choices? Hardly. There is a gas line or there isn't and one can look for alternate sources of supply of natural gas about as successfully as one can seek an alternative telephone company. Much natural gas is distributed by the public utilities that also distribute electricity. It would be dangerous and foolish to give up some public check on this activity. The best policy may be to raise the price of natural gas in the interest of preserving supplies for premium uses. But in no case should such a decision be left to private corporations.

The energy industry has enjoyed a favored tax position for many years. Depletion allowances permit oil companies to deduct 22 percent of

Table 11-3.   Corporate Income Tax Paid by Five Largest Oil Companies

Company	1971			1962-1971		
	Before Tax Profit (billions of dollars)	Percent Tax Paid	Approx. Loss to Treasury (billions of dollars)	Before Tax Profit (billions of dollars)	Percent Tax Paid	Approx. Loss to Treasury (billions of dollars)
Standard Oil (N.J.)	2.74	7.7	0.885	19.65	7.3	6.4
Texaco	1.32	2.3	0.495	8.70	2.6	3.3
Gulf	1.32	2.3	0.500	7.86	4.7	2.8
Mobil	1.15	7.4	0.375	6.39	6.1	2.2
Standard (Calif.)	0.87	1.6	0.330	5.19	2.7	1.9

Source: *Progressive* magazine

gross receipts, not to exceed 50 percent of income. Besides this best known of the tax subsidies to the industry, there are other provisions in current tax laws permitting liberal write-off of capital expenses, deduction of certain indirect costs as current production costs (even though other industries must count similar expenditures as capital costs), deduction of foreign operating costs even if the operation failed to show a profit, and a variety of capital gains shelters. Large amounts of money are involved in these subsidies. One effect is the encouragement of exploitation, discouragement of draining oil and gas fields, and what has been called the "drain America first" policy. But what concerns us here is the failure of the market and the distortion of energy costs. Table 11-3 shows the taxes actually paid by the five largest oil companies. These are typical values for the oil industry. In 1969, 1970 and 1971 the average corporate income tax paid by the oil industry was 5.8 percent, 10.1 percent and 6.1 percent.[4] This is to be compared to an average of about 42 percent for all industry.

Table 11-3 shows that, in 1971, these oil companies failed to pay their share of taxes in the amount of about 2.6 billion dollars, and thus, each person paid about $12.93 more in taxes than he would have otherwise. For a family of four, this would amount to about 10 percent of their annual costs for fuel and energy. If tax subsidies to the rest of the industry are included, we estimate that each citizen paid an amount in taxes equal to about 20 percent of his energy costs. Without these subsidies, fuel costs would no doubt be higher, taxes on individuals lower, and the consumer would choose how much fuel to buy at higher prices; this situation would come closer to a free market. The API is aware of that, and it makes their appeal to the free market ring hollow.

The reasons given by the energy companies for needing these public subsidies are that the business is risky, incentives are needed for exploration and development, and incentives are needed to permit accumulation of capital. If the business were risky, one would expect to find large fluctuations in profits, periodic business failures and the rise of new companies as fortune smiled first on one and then another. These conditions do not obtain. From the reports of Council of Economic Advisors to the Congress, it is clear that fluctuations in profits are about the same in the oil business as in other businesses. Further, although a few companies have made a successful start in the oil business in the last twenty years, they are very few indeed, and no large companies have gone under. The fact is that the risks are well understood, estimated, and shared. For example, almost all the offshore operations (which are among the most risky) are underwritten by groups of companies so that a single failure means little. Finally, if the business were really as risky as the energy companies sometimes imply, featuring periodic giant profits and punctuated by disastrous losses, they would have advanced the most powerful argument for direct government control or operation, so that this essential supply could be assured.

Profits have been high in the oil industry for years. In refining, for example, profits after taxes have averaged 10.8 percent for the last twenty years. By comparison, the average for all industry for the same twenty years is 5.1 percent after taxes.[5] The energy shortages that began to be felt in 1972 may have caused inconveniences, but have not damaged the profits of the oil companies. Profit increases for the first half of 1973 ranged from a "low" of 37 percent for Standard of Indiana to a staggering 174 percent for Commonwealth Oil.

Walter Heller, former Chairman of the Council of Economic Advisors, sums up the position of most experts outside the oil industry:

> To continue stimulating the overexploitation of oil by tax subsidies in the form of excessive depletion allowances, capital gains shelters, and special deductions becomes ever more anomalous. Here is another case where believers in the market-pricing system ought to live by it. The public is subsidizing these industries at least twice—once by rich tax bounties and once by cost-free or below-cost discharge of waste and heat. . . . The net result is to underprice and overproduce petroleum products and the energy derived from them.[6]

It is certainly true that any collection of energy options for the future will require accumulation of capital, probably large amounts of it. The strange implication in the API statement—and in the statements of the individual oil companies and the electric utilities—is that the industry ought to, or must, finance this sort of expansion from current earnings. This view denies the possibility of investors participating in the growth of industry by purchasing stocks or bonds. It suggests that the energy industry should not—or cannot—compete in the money markets of the nation as even the government must do. As a practical matter, it would further concentrate the control of energy resources and possibly limit the options for the future.

(3) A third set of recommendations from API relates to import policy. They call for import restrictions to prevent free entry of foreign oil and gas. The best clue to the philosophy behind this call is contained in the statement that "the import system should be unencumbered by objectives unrelated to national security." This sounds strangely like the "fortress America" policy that proved so disastrous before World War II. By national security, the API makes clear, they mean a domestic supply of oil. But national security could better be had by reducing world tensions or by reducing dependence on oil.

Several of the recommendations on imports were contradicted by public needs and energy shortages, and import quotas were largely abandoned in April, 1973—as had been recommended by a presidential task force four years earlier.

(4) Recommendations for synthetic fuels are similar to the recommendations for regulations and taxes. More government (taxpayer) support is recommended in the form of research support and tax subsidies. Every taxpayer should ask himself whether he wants to supply gifts of money and guarantees of profit without the establishment of adequate controls.

(5) Recommendations for environmental protection again demonstrate a parochial point of view. API suggests that "government has the responsibility to establish realistic environmental standards in close consultation with industry." There are many other groups that government should consult as well, but API finds no place for them.

The National Petroleum Council—an oil industry group set up by the Secretary of Interior to advise him—is less shy. They call for the government "to eliminate the serious delays that have been caused by environmental issues."[7] To the extent that these delays are the normal time taken to resolve conflicting goals in the courts, they are part of the price of a pluralistic society.

But there is a clear implication that the proximate cause of our energy problems is the delay caused by opposition to the plans of the energy industry on environmental grounds. Among themselves, oilmen are more candid. Robert Buschman, president of the Texas Mid-Continent Oil and Gas Association, described at the association's 1972 meeting the "shortsighted environmental policies" as one of the "ill-advised — if not downright — foolish federal policies."[8]

When environmental opposition halts construction of a power plant it is news. When other kinds of delays occur it is not news. An accounting of delayed projects suggests that the industry's attempt to blame delays on environmentalists is simply self-serving. For example, 72 percent of the electric power plants that came into service between 1966 and 1970 experienced delays in construction. Fifty-two percent of these delays were caused by labor problems, 23 percent by equipment failure, 14 percent by late delivery of equipment, 6 percent by slowness in regulatory clearance—including, among other things, environmental opposition—, and 5 percent by weather and design changes. Several plants were delayed for more than one reason. Of the 230 power plants scheduled for completion between 1971 and 1977, 53 were reported as delayed (in 1972). Of these, 30 were being delayed by labor problems, 27 by regulatory matters (including, but not limited to, environmental opposition), and 12 by weather, design changes, and other non-environmental matters. (Multiple causes were reported for some delays.)[9] Of course there are cases when environmental considerations and energy industry plans come into direct conflict, but the industry will have to look elsewhere for a villain to take most of the blame.

(6) API's final recommendations suggest coordination of federal

energy programs toward a long-range energy policy. This goal is certainly desirable. We indicated earlier in the chapter that a long-range policy is necessary if we are to maintain a supply of energy sufficient to our needs. Government energy policy must have provisions for meeting our needs, developing alternative energy sources, distributing energy among the various sectors, and reducing consumption of energy.

Before proceeding to a discussion of energy policy, we should examine the agendas of the groups that have an interest in that policy. Each of these groups has its own priorities and needs. Their interests may be conflicting, and energy policy must be carefully formulated with the aim of reconciling them. These parties include industry, the energy-consuming public, government, the electric utilities, and other countries.

Industry is one of the largest energy consumers. It is committed, of course, to survival and, almost without exception, to continued expansion. For these ends, a cheap and reliable supply of energy bulks large in fundamental plans. However, the recent energy shortages have demonstrated that the era of cheap energy supplies is over and reliable supplies are in doubt.

The energy shortage threatens continued industrial growth. The desirability of such growth can be questioned, but industrial leaders and many economists (Walter Heller, James Tobin, William Nordhaus, and Robert Lampman, for example) believe that social improvement will end if growth rates are reduced and we eventually arrive at a steady-state economy. The argument goes like this: we have had social gains and improvements in the material standard of living in this period of growth. Since growth has accompanied these changes, the end of growth would mean the end of social improvement. Among the evidence cited is the correlation between increases in the Gross National Product and energy use and the improvement in social conditions. Counterarguments by two economists who have studied the growth issue in great detail, E. J. Mishan and Nicholas Georgescu-Roegen, point out the uncounted costs of economic growth and the theoretical reasons why growth must stop.[10] We have already seen (chapter 3) that exponential growth in a finite world must eventually stop. What is not yet known is when and how growth will slow or end. These matters are complex, and serious discussion of them has only recently begun.[11] For the present, industrial leaders can be expected to press for energy to continue expansion of industrial output.

In support of the basic aims for survival and continued expansion, industry will probably urge relaxation of environmental standards, expansionist monetary and fiscal policy (even at the expense of inflation), and high priority for industrial fuel allocations.

The energy-consuming public includes all of us. We have under our direct control the two largest end uses of energy—the automobile and home heating (see chapter 9). It is not surprising that the first federal policies for energy conservation were directed at these two uses.

The goals listed at the beginning of this chapter are in the interest of every citizen. Unfortunately, few people seem to understand the ways in which these goals are related to energy policy. This lack of public understanding is being compounded by the persistent presentation of the points of view of industry and government in the public media. While these presentations are usually true as far as they go, many options do not get mentioned.

The only expression of grass roots opinion on energy-related matters that has resulted in legislation in the last few years is in the environmental protection area. It will require some public vigilance to insure that all environmental gains are not sacrificed in the name of energy shortages. It is worth recalling that much of industry, the AEC, and most of the energy companies actively opposed the National Environmental Policy Act, or sought to exempt themselves from it by special provision.

But the environment is only part of the story. Somehow the public must learn that issues like the highway program, mass transit, zoning and building codes, and public expenditure on research and development will determine the future of energy supply and the fabric of their own lives.

Although we have a representative government, governmental policies do not necessarily reflect the desires of the public. Governments, like many large institutions, acquire an agenda and momentum of their own.

Historically, the energy industry has received generally favorable treatment from the government. Things have not changed much in the recent past. For example, the special energy program outlined by President Nixon on April 18, 1973, corresponds almost point-by-point with the recommendations of the American Petroleum Institute. Only on the matter of oil import quotas did the administration go against oil industry policy. The fuel shortages that occurred in the winter of 1972-73 impelled the president to end the restrictions on oil imports.

Government energy policy depends partly on history, partly on the relative strength of pressure groups, partly on public opinion, and partly on the options that are perceived. Unfortunately, it is made almost always for short- or intermediate-term goals. The following long-standing policy directions have considerable effect on the government's reaction to energy issues:

(1) Pursuit of continued exponential growth in the economy. Although we know that such growth must eventually stop, there is no hint of any policy anticipating this certainty. In fact, it seems that government energy policy will be directed toward satisfying the demand for more energy. It is unlikely that government will seek to slow the growth rate in an effort to reduce energy consumption.

(2) Commitment to nuclear energy. The early spectacular successes with nuclear weapons technology and optimistic estimates of the future of nuclear power have produced a well-established bureaucracy in the executive branch and in Congress. Despite recent conflicts over nuclear power, these vested interests can be expected to defend the primacy of a nuclear future against all alternatives. New voices are being heard on this subject, but so far there has been little change. Consider, for example, the funding of energy research and development: for the 1969-1973 period, three-fourths of all appropriations were for nuclear sources, with the largest increases going to fusion research. Should it appear desirable to diversify, these vested bureaucratic interests will pose a problem.

(3) Fears for national security. This highly emotional issue has been intensified anew by the use of fuel supplies as a policy threat by the Arab nations. The resulting policy pressure has been for energy self-sufficiency as soon as possible. While energy self-sufficiency may be a desirable goal, it will not produce national security. Neither the world nor the United States can be secure as long as western Europe and Japan are almost entirely dependent on imported fuels.

(4) A marked preference for technological rather than institutional or social solutions to energy problems. Despite a general awareness that marked technological changes introduce large and unpredictable social changes, the government has an unvarying record of seeking technological and engineering solutions to our problems. If the public should decide against an all-nuclear energy future, they will have to make these wishes known in spite of government pressure.

The electric utility industry has long been organized as a lobby in Washington and in the state capitals. Its lobbying effectiveness has been strengthened by strong public reaction to even brief interruptions of electric service. In fact, the Public Service Commissions of the states, feeling powerless to solve the larger problems of regional supply and distribution of electricity, have importuned the federal government to take action in this area.

In the past twenty-five years, the electric utility industry has seen rapid and continuous growth, accompanied by declining unit costs. Large users have been offered progressively lower rates for electric service on the grounds that the reduced unit costs meant lowered rates for all users. But we seem to have entered a new period in which further expansion no longer brings declining unit costs. Higher fuel costs and cost overruns in equipment for nuclear power plants will further increase the unit cost. Under such circumstances, these promotional policies and rate structures are no longer defensible. Declining rate block structures mean that the low volume users—who pay the highest rates—are subsidizing industry's expansion through ever higher rates.

The manager of an electric utility is in a peculiar position, however.

Because user rates, rate of return on investment, and many other features of his operation are subject to public control, his only way of increasing profits is through expansion. Thus we can expect the utilities to push for an expansionary energy policy, even if the environment must suffer and small users must subsidize large users. The usual utility argument that it is cheaper to service large users than small users is self-serving. If rates were to be made solely on a cost-of-service basis, then they should also be a function of the user's distance from the generating station. Long ago it was decided that such a charge was inadvisable. The point is that we already violate the cost-of-service principle.

Thus the electric industry is likely to favor low fuel costs, expansionary policies, and reduced opposition from environmental groups. If energy shortages worsen, the electric utilities will seek relaxed environmental standards for utility construction and operation.

For many years the United States and other industrialized nations have relied on relatively cheap supplies of imported fuels and materials which they needed. The 20 percent of the world population in industrialized nations have for many years consumed about 80 percent of the depletable resources produced each year. Furthermore, these supplies were often produced by U.S. companies, or those of other industrialized nations operating under foreign concessions. This state of affairs was justified—even made a source of pride in colonial days—on the grounds that such resource production added to the economies of the poor countries that were the sources. The low prices were justified on the grounds that the free marketplace was functioning.

But things are changing in the poor nations of the world. Their governments have realized that they were parting with their resources very cheaply. In an inversion of the old roles, the lesser developed countries with large unexploited resources play off the industrialized nations against one another for the best deal, much in the way that the rich nations once shopped among the poor nations for oil and minerals.

In the case of petroleum fuels, the poor nations have banded together to obtain higher prices. The Organization of Petroleum Exporting Countries (OPEC) has become a powerful force in the world search for assured fuel supplies. Should we oppose the OPEC grouping as a monopoly? If it is a monopoly, it is hard to know just what we could do about it that would not worsen world tensions.

The practical course would be to vary our energy supply options. If our energy resources were diversified, we would not feel so threatened by a cutoff of our Middle East oil supply.

Meanwhile, if the OPEC nations should nationalize their petroleum resources, the United States must react with caution. If these nations seek to purchase control of foreign oil companies' operations within their borders, their demands should be acceded to gracefully—for the alternative is seizure without payment. After nationalization is accomplished, these countries will seek to reenter the international market to sell their oil. As the world's largest customer, we will have to exercise discretion to prevent ourselves from taking unfair advantage at that time.

For the OPEC nations, nationalization of oil resources may be the only course to a better future. It is the only means whereby they can achieve an improved standard of living for their populations, and it behooves the United States to help poor nations in this effort.

We *do* have an energy policy in the United States, but most of the practices that make up this policy were not thought out as part of a consistent program. Nevertheless, the conglomeration of activities, regulations, restrictions, and plans for energy supply must be the point from which any attempt at a coherent policy must start.

While energy was cheap, environmental effects of its use slight, and fuel shortages not troublesome, most users had a choice of energy sources. Interfuel competition, however, applied primarily to initial choices. Once an industrial process was chosen or a furnace installed, the user was committed to a single fuel. Choice was further modified by technology—for example, steam and electric automobiles disappeared many years ago, and gasoline is the only fuel choice for current autos.

Government subsidies to some fuels further modify the market. Environmental concerns and public expenditures—on roads, for example — limit or distort the choice of fuels still more. Persistent fuel shortages will bring further limitations.

We cannot recover an unmodified free market in fuel supply any more than we can return to the days when people cut their winter fuel supply from the land on which they lived. We could attempt to return to the free market, but it would be a long and tortuous path, and we should consider the reasons why society intervened in the first place. There were often good reasons, whether or not the intervention solved the difficulties.

Scattered among sixteen agencies—or forty-two, depending on who is counting—are a wide variety of regulatory functions of the federal government. Among the most prominent of these agencies are the Federal Power Commission (FPC), the Interstate Commerce Commission (ICC), the Atomic Energy Commission (AEC), the Environmental Protection Agency (EPA), the Civil Aeronautics Board (CAB), the Federal Aviation Administration (FAA), the Bureau of Mines, and the Oil and Gas Division of the U.S. Geological Survey. But even these agencies are only a part of the federal activities that help determine our present energy policy. All agencies that are active in the transportation area influence fuel and energy use. Other agencies involved with housing, agriculture, health, land use, transportation and shipping, and a host of other government activities directly or indirectly influence energy sources and uses.

With the advent of fuel shortages in the early 1970s, hearings, legislation, special task forces, and consultants appeared everywhere. Because energy is one of the two fundamental inputs to an industrial society, decisions of many kinds, from mining safety laws to FHA mortgage conditions, have impact on energy supply or consumption. There is simply no realistic possibility of combining all energy-related activities in one agency of the government—state or federal. One might as well try to bring together in one agency all activities that require typewriters.

It is clear, however, that until recently the various agencies have paid little attention to their common relationship to energy supply and consumption. It is this lack of coordination that many have in mind when they conclude that the United States has no energy policy. We share in the basic dilemma of all industrialized countries. First, we have a great many problems and objectives which frequently come into conflict when the making and execution of policy is considered. Second, though there may be an optimum organizational structure for each particular problem, we can have only one organization at a time — and whatever one we have must be a compromise that attempts to deal with a variety of problems simultaneously. As society becomes more complex and its problems become more numerous and serious, an organizational structure's effectiveness in dealing with any given problem will decrease. Sloganeering of the sort the energy companies and some politicians engage in will not help under these circumstances.

For better or worse, the government does intervene in the workings of the energy supply and consumption activities of society. It is almost certain that we will continue to regulate, license, support research, and in other ways influence the use and availability of energy. Of course these activities should have better coordination. To that end, structures like a Council of Energy Advisors deserve attention, but the limitations of such arrangements should be realized. Public discussion of options and the preservation of a wide variety of options are often the result of conflicting views. Centralization of authority will not encourage public discussion. Furthermore, the choice of energy policy certainly involves decisions about employment of the nation's natural resources (public and private), foreign policy, environmental standards, and, at times, social objectives, economic stability, conditions for workers, and the very survival of our society. Such decisions do not belong in a coordinating group. No matter how messy it may be at times, these decisions belong to the people of the nation engaged in the political process.

Regulation also will continue, even if we should decide tomorrow to reduce it as far as possible. The difficulty of regulating a technically independent industry must still concern us. An astronomer who was serving on an advisory board of the FCC once remarked that one shouldn't stay in that kind of activity too long because sooner or later it is easy to become sympathetic with the problems of the regulated industries. That has been a persistent problem in regulatory agencies. The regulators, in the course of learning their work, come to understand that the problems of the regulated are real. When the regulated activity is also a technical one—as is the case in matters related to energy—the principal source of qualified people is the industry itself. The solution to this difficulty is not to be found in a search for unbiased men. It is likely that when you find someone who is unbiased about the oil industry, you will have also found someone that knows little or nothing about it. Such a man is an easier mark for the manipulator in industry than a person who understands oil—biases and all.

Since we are committed to continued regulation, it may be time to recognize regulatory activity as a permanent feature of all industrialized societies. Regulatory actions have more in common with judicial proceedings than with the administrative ones with which they are ordinarily grouped. The time may have come to group all regulatory activities together, and separate them from both the executive and judicial branches of government. Then the government may be able to build up the kind of procedural safeguards that will permit the regulators to avoid becoming the captives of the regulated. In the case of energy regulatory activities, the grouping together of standard setting, research and development, promotional activities, and regulation was tried with the AEC. This arrangement has an inherent conflict of interest in the assigned activities, and has made the AEC subject to considerable — inevitable but often undeserved—suspicion. David Lilienthal, first chairman of the AEC, concluded that the arrangement did not work, and has long advocated the separation of AEC activities.[12]

It has been pointed out that our energy problems do not yet represent an absolute resource shortage. We could, if we chose, mount a massive effort to supply our present needs and those of the next few decades through conventional sources. But a choice for massive exploitation of traditional fuels—besides having a fearful price—will only deplete these resources all the sooner. We would then have the same problems with a smaller base of resources to solve them.

Choosing policies for short-term advantage could be disastrous if the long-term results are ignored. When considering our options—and we can cover only a few of the possibilities here — we must remember that future generations will have to live with the consequences of policies chosen in the next few years.

A coordinating and advisory body is needed at the level of the Executive Office of the President. It could be a statutory group, one set up by executive order, or one located within the existing structure (for example, within the Office of Management and Budget), but it should be required to report periodically to the Congress and the public on the state of energy resources, research, achievement of policy goals, and the like. A Department of Energy and Resources could be useful—and there is, in Canada, an example of this arrangement worthy of study—but much energy-related activity has been handled by the Department of Interior, and that has not solved our problems.

The most difficult organizational problems arise in connection with regulatory activities, which must be well separated from the action programs of government. The possible options range from a facelifting of present regulatory agencies to the establishment of new, separate regulatory agencies, to the more dramatic step of instituting a separate branch of government, coequal with the present branches, which would implement all regulatory efforts, including those for energy matters. With several different kinds of regulatory activities proceeding under one set of leaders, it might be possible to construct the kind of procedural safeguards that have characterized the courts, including the right of appeal.

In the Congress, progress on energy matters depends upon finding ways to change the jealously guarded fiefdoms of the congressional committees. At present, more than twelve committees claim jurisdiction over various energy issues. These jurisdictions often overlap. As a practical matter, it might be possible to abolish the Joint Atomic Energy Committee and replace it with a Joint Energy Committee. This is a minimum measure. More desirable measures will have to wait for resolution of some of Congress's longstanding archaic practices. Congressional reform has been sought for many years, but with little success.

Litigation surrounding power plant siting has revealed a longstanding institutional shortcoming of our court system. With opposing litigants presenting expert testimony on matters of considerable complexity, the normal protagonist procedures of the court are inadequate. The courts need some means of obtaining third-party expert information. Such a procedure would frequently bring evidence into the court which is widely known in the technical community—and relevant to the question at issue—but not introduced by the contending parties for reasons of their own.

These institutional suggestions have been made many times. If we are really to seek energy for policy, we must recall the goals of society and seek the institutional changes conducive to achieving those goals. One of the clearest messages heralded by the energy shortage is that change is coming.

We should seriously consider direct public control of fuel resources and major facilities for energy conversion. There is considerable experience with varying degrees of public control in many countries, including the socialist countries. The lessons from these arrangements in times of energy shortage are not yet clear, but it appears that central control is no assurance against the development of energy shortages nor does it guarantee efficiency.

The record of government operation in the United States is more encouraging. The TVA has its weaknesses, but it is a success by nearly any standards in a program that would have been hard to execute with a private industrial operation. Similarly, the many municipal utilities show no particular disadvantage compared to privately owned utilities. Perhaps we should set up several large public energy corporations, modeled on TVA, to test the success of public operation of energy supply and to provide a standard against which the performance of the energy companies can be measured. We think it is at least time to apply public utility control to the energy industry. But our own mythology gets in the way: the word nationalization has evil connotations to many.

In addition, legislative measures can be instituted to bring a reduction in energy use without changing basic living patterns, and without aiming to lessen growth. But such measures can only buy a few more years. Some time to solve our problems would be welcome, *if* the time is actually employed to seek solutions, and not—as is usually the case—simply used to forget the basic problem for a while.

Measures could be taken to accelerate supply and production of presently used fuels and energy. But this would be foolish for the production of fossil fuels, since the handwriting of change is already on the wall. Expanding fossil fuel production would be justified only as a part of a program to seek new energy resources, with the fossil fuels used to alleviate short-term shortages.

The most promising strategy would be to mount a variety of programs designed to produce energy from a wide range of sources, especially on-site sources, such as solar home heating, and to institute research into those sources which still must be proven feasible. Measures to encourage less use of energy (such as tax incentives to individuals) could be very helpful. Prohibitory measures should be taken with caution, but it may be necessary to limit the size of automobiles, since it appears that appeals to the auto industry have but limited effect.

For regulatory agencies, it is time to consider using the rate-making power as part of an overall energy policy. Recent studies indicate that the elasticity of the demand for electricity is considerable, and it would thus be possible to use the rate structure to inhibit growth in demand. At least it is time to end promotional practices and rate structures, unless it can be proven that such practices will bring a reduction in unit cost.

The United States is heading toward commitment to a nuclear future, but the option for more varied approaches is still open. The vision of a nuclear utopia has long since been discredited, and the director of Oak Ridge National Laboratory now calls it (approvingly) "a Faustian bargain."

We have discussed many alternative energy sources. The public image of many of these sources is somewhat distorted. The public has heard that breeder reactors are almost here—and almost safe—and that other energy sources are futuristic stuff. Meanwhile, amateurs are heating hot water and homes with local solar energy conversion, farmers are constructing methane generators which operate on animal wastes, hot steam geothermal generation of electricity is succeeding in California and Italy, and modern windmills are doing very well in Australia and Denmark. None of these sources is any more likely to solve all our problems than is the breeder reactor, but the technology is understood and operating. The principal problems are with cost and engineering, which is quite a different thing from MHD, thermionic, or thermo-electric power generation where much basic work remains to be done.

For those energy sources whose principal problems are cost and engineering design, the space program is a model. The space program was an engineering success. Consider where we might be in energy development if we invested billions of dollars each year for a decade on solar energy. If that is unconvincing, consider the reverse case. Where would we be in space if we had spent a million or two each year in the 1960s?

In the end it comes down to what we want. Until goals are established, planning for the future is not possible. It becomes, in the words of the OECD planning conference in Bellagio, Italy, "an effort to make the inherently bad . . . more efficiently bad."[13] The United States has a strong technical base. When we have the goal, the energy limitations on it, if any, can be discovered. If there are none, the goal can be met.

References

1. B. C. Netschert, "The Energy Company: A Monopoly Trend in the Energy Markets," *Science and Public Affairs* 27 (1971): pp. 13-17.
2. *Ibid.*, pp. 13-17.
3. American Petroleum Institute, *Statement of Policy: Energy* (New York: November 1972).
4. *Congressional Record*, July 19, 1972, H6707-16.
5. *Economic Report of the President*, prepared by the Council of Economic Advisors, various years.
6. W. W. Heller, "Coming to Terms with Growth and the Environment," in *Energy, Economic Growth, and the Environment*, ed. S. H. Schurr (Baltimore: Johns Hopkins Press, 1972), pp. 3-29.
7. National Petroleum Council, *U.S. Energy Outlook* (Washington, D.C., 1972).
8. R. A. Buschman, "Presidential Address" (Address delivered at the Texas Mid-Continent Oil and Gas Association, Houston, Texas, October 10, 1972, mimeographed.
9. J. A. Lieberman, in *Energy and Public Policy — 1972* (New York: The Conference Board, 1972), p. 14.

10.   E. J. Mishan, *The Costs of Economic Growth* (New York: Praeger, 1967); N. Georgescu-Roegen, *The Entropy Law and the Economic Process* (Cambridge, Mass.: Harvard University Press, 1971).
11.   American Academy of Arts and Sciences, "The No-Growth Society," *Daedelus,* Fall 1973.
12.   D. Lilienthal, *Change, Hope and the Bomb* (Princeton, N.J.: Princeton University Press, 1963).
13.   Quoted in E. Jantsch, *Technological Planning and Social Futures* (New York: John Wiley & Sons, 1972), p. 256.

12

RETURN TO THE FUTURE

Upon this gifted age, in its dark hour,
Rains from the sky a meteoric shower
Of facts . . . they lie unquestioned, uncombined.
Wisdom enough to leech us of our ill
Is daily spun; but there exists no loom
To weave it into fabric. . . .

Edna St. Vincent Millay

A man, viewed as a behaving system, is quite simple. The apparent
complexity of his behavior over time is largely a reflection of the
complexity of the environment in which he finds himself.

Herbert Simon

If we are to take any hand in the future, to aim at any goal, we must first recognize the patterns imposed on us by the present and the past. If we wish to know how much we need of energy or resources, we should first understand in which direction we are going.

At present, the universal aim is continued growth. Rich and poor nations alike strive for increased industrialization, more goods, more specialization, and—to drive the whole process—more energy. One would suppose that this universal aim of nations implied great satisfaction with the course of history, seeing it as a steady improvement in the human condition.

241

The largest monuments left to us by ancient Egyptian civilization were the great pyramids. The largest monuments of American civilization may be its sprawling highway interchanges. This is the junction of Interstate 10 and Interstate 405 in West Los Angeles. (Photo courtesy of Division of Highways, California Department of Public Works.)

But is the present better than the past? In chapter 2 we indicated that evidence on this point is mixed. Many people are better off than ever before, living in conditions of material abundance that were available only to monarchs and priests a few centuries ago. But world population has increased so vastly that there are also more people — larger absolute numbers of persons — living in abject poverty than at any time in the past. But since many are living in relative freedom from hunger and disease and fear, we view history, especially recent history, as progress.

The continuing growth and industrialization of modern times has generated a litany sanctioning inequalities and injustices even while taking pride in the increased numbers of the well-to-do. The goal of industrial society was to have enough for everyone — to have freedom from hunger, disease, fear; and an opportunity for each individual to search for satisfaction.

Most industrialized countries can now achieve these goals, but although inequalities continue, they show no signs of slackening their growth or sharing their excess with the poor of the world. In some respects, the rich industrialized nations have achieved their growth at the expense of the poor nations.

A group of distinguished British scientists summarizes the problem this way:

> The principal defect of the industrial way of life with its ethos of expansion is that it is not sustainable. Its termination within the lifetime of someone born today is inevitable—unless it continues to be sustained for a while longer by an entrenched minority at the cost of imposing great suffering on the rest of mankind. [1]

We shall have growth for a while yet, perhaps for a long while in the poor nations of the world, but the industrialized nations must prepare to end the kind of growth that is achieved through use of ever more energy. We will not consider how the privileged few might maintain their position against the destitute many.

The question at hand is energy. We need to know the minimum amount needed and how to get it, and how much more can be counted on as a long-range supply. If our civilization is to be more than a brief and aberrant experiment among civilizations, this long-range supply must not encounter obvious limits for at least a millennium.

As the public media began giving increased attention to fuel shortages and the energy crisis, graphs showing the amount of energy used in the recent past began to appear. On these same graphs were projections showing how much energy their author thought we would use in the future. The simplest of these projections are just what they appear to be — extrapolation of past totals into the future. But such extrapolation is guaranteed to fail when past trends and relationships change, or take new directions. This basic weakness constitutes an advantage if the basic forces causing growth are not understood, and do not change. In such cases, gross projections of totals may be more accurate than more complex methods.

Most projections of energy use are made from models of some sort. The most common type of projection divides energy use (or supply) into a number of categories and sectors and projects each of these separately. Totals are obtained by recombining the categorical estimates. In this approach—sometimes called the building block approach — assumptions are made about the behavior of variables that are assumed to effect the level of future energy use. Among these variables are increases in gross national product and population, fuel prices in relation to prices of other commodities, availability of fuels, and technological change. Even with this small number of variables, models of considerable complexity can result, and there may be difficult questions about the quantitative behavior of the variables considered.

Figure 12-1 presents twenty-six estimates of energy use forecast for the year 1980. The estimates are from many sources and represent a wide variety of methods and models. If the estimates are separated and compared according to the year in which they were made, none differs from other contemporary forecasts by as much as 100 percent. The trend in recent years toward higher values for the 1980 estimates is clearly seen. With one exception, forecasts made as recently as 1972 show nearly as much scatter as ones made earlier. This probably reflects differences in the assumed effect of related variables and their future magnitude. In any case, comparison of actual energy use with the forecasts strongly suggests that, despite the method used, earlier forecasts reflect the usage of the late 1950s and early 1960s, and later forecasts strongly reflect the increased energy use experienced in the late 1960s.

One forecast in figure 12-1 predicts a range of possible values for 1980. This is not a measure of uncertainty in the usual sense. Such a forecast is a contingency forecast, that is, one which tests several values of the related variables (GNP growth rates, technology changes, population growth, and so on) and estimates the results of the various combinations. Such estimates are sometimes requested by policy makers to assess the effects of various policy changes.

Figure 12-2 presents thirteen forecasts of energy use for the year 2000. As might be expected, these forecasts vary more than do those for 1980. The trend toward higher estimates reflects the recent acceleration in energy use. This suggests that—as in the forecasts for 1980—recent experience is the most heavily weighted.

The matter of estimates of future consumption cannot be treated here with the depth needed. At present, extensive efforts are underway to provide more reliable estimates. These efforts should be encouraged, but with the understanding that forecasts can become self-fulfilling prophecies. A considerable amount of skepticism is warranted, even required, until the conceptual problems are solved, if indeed they ever can be. Meanwhile, the work should continue, for one cause of the shortage of electrical generating capacities in the early 1970s was that future needs were underestimated at the time new capacity was planned.

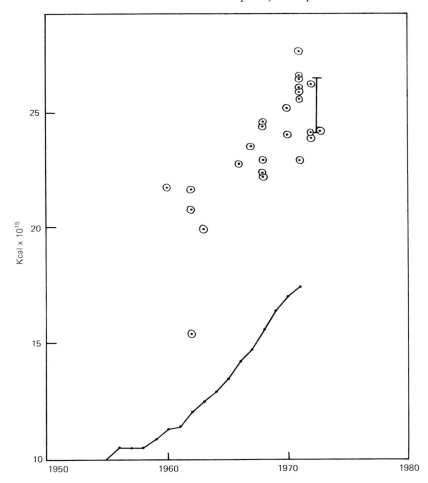

Figure 12-1. Published forecasts of energy use. Twenty-six forecasts of U.S. energy use for 1980, compared with actual energy use. Forecasts are plotted according to their year of publication. Sources: U.S., Congress, House, Committee on Interior and Insular Affairs, *Energy "Demand" Studies: An Analysis and Appraisal* (Washington, D.C.: 1972); Battelle Memorial Institute, *A Review and Comparison of Selected United States Energy Forecasts* (Washington, D.C.: 1969); U.S., Department of the Interior, *United States Energy Through the Year 2000* (Washington, D.C.: 1972); Shell Oil Company, *The U.S. Energy Position* (Shell Oil Co., 1971).

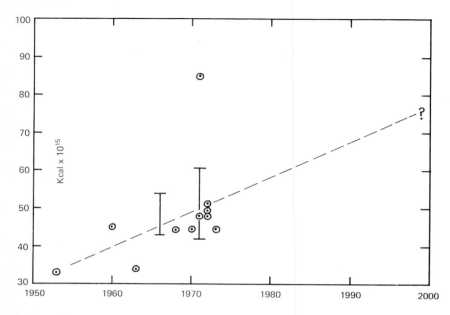

Figure 12-2. Thirteen forecasts of U.S. energy use for the year 2000.
Forecasts are plotted according to their year of publication. Trend line is
suggestive only. Sources: U.S., Congress, House, Committee on Interior and
Insular Affairs, *Energy "Demand" Studies: An Analysis and Appraisal*
(Washington, D.C.: 1972); Battelle Memorial Institute, *A Review and Com-
parison of Selected United States Energy Forecasts* (Washington, D.C.: 1969);
U.S., Department of the Interior, *United States Energy Through the Year 2000*
(Washington, D.C.: 1972); Shell Oil Company, *The U.S. Energy Position* (Shell
Oil., 1971).

Throughout this book we have concentrated on fuels and energy
supply for the United States. But we in the United States cannot operate as
if the remainder of the world will fit smoothly into whatever plans we
choose. Policies of other nations are of crucial importance to our fuel
supply (and for our supply of other mineral resources as well). Internal
policies of other nations are made with their own needs in mind, not ours,
and to begin to estimate the future for the United States requires a hard
look at the future situation for the world. Figure 12-3 shows very clearly
that changes are in store. Without projections of any sort it is clear that the
past course of development must and will be altered. Even the optimistic
estimates in figure 12-3 demonstrate that fossil fuels, which have supplied
the bulk of the world's energy for three generations, are inadequate to
supply the world's needs much beyond the middle of the next century. But
projections of past experience assume that the present imbalances in distribution of goods and services which leave 80 percent of the world's pop-

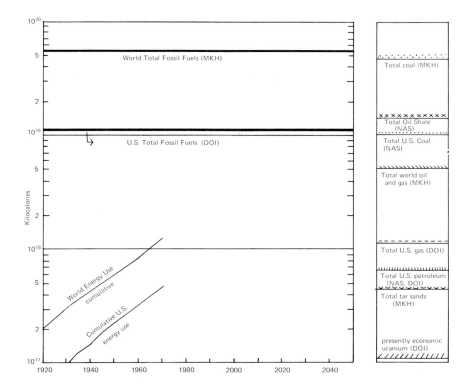

DOI: U.S. Department of the Interior estimate

MKH: M. K. Hubbert estimate

NAS: National Academy of Science estimate

Figure 12-3. Cumulative world and U.S. energy use compared to total fossil fuel resources and totals of some of the component resources. These estimates include submarginal resources not now exploitable. Note that nearly 90 percent of total fossil fuel resources are coal. Sources: U.S. Department of the Interior, *United States Energy, A Summary Review* (Washington, D.C.: 1972); M. K. Hubbert, "Energy Resources for Power Production," in *International Atomic Energy Symposium on Environmental Aspects of Nuclear Power Stations* (Vienna: 1971); National Academy of Sciences, *Man, Materials, Environment: A Report to the National Commission on Materials Policy* (Washington, D.C.: 1973). Other sources: U.S. energy use: S. H. Schurr and B. C. Netschert, *Energy in the American Economy, 1850-1975* (Baltimore: Johns Hopkins Press, 1960); U.S. Department of the Interior, *United States Energy Through the Year 2000* (Washington, D.C.: 1972); world energy use: J. Darmstadter et al., *Energy in the World Economy* (Baltimore: Johns Hopkins Press, 1971).

ulation at poverty level will continue. The poor nations of the world are not willing to accept such an assumption. If allowance is made for their attempts at rapid improvement in the material standard of living for their citizens, and the energy which that will take, the crunch for energy becomes even clearer.

It is easy enough to declare an absolute per capita minimum for energy requirements. For thousands of years people lived at a personal energy consumption level of 2500 to 3000 kilocalories per day—the energy equivalent of the food they hunted and gathered. With the advent of agriculture and the first domestic animals, per capita energy use grew from this level to 10,000 to 15,000 kilocalories per day, where it remained until well into the Middle Ages. But these limits are of little interest today—although much of the world has yet to exceed them—because we cannot, at present population levels, recover those conditions even if we should wish to do so.

We know of only one effort to estimate the minimum energy necessary to provide a satisfactory life. Everett Hafner has produced such an estimate.[2] For production of food, 300 watts' continuous expenditure is allowed, or about 6200 kilocalories per day. In the light of the food-energy relations described in chapter 4, this energy subsidy would be approximated by 1920s levels in the United States food system, but in many other nations this level would represent a considerable increase of the present energy subsidy to food production.

The same amount, 6200 kilocalories per day, is allowed for shelter. Hafner assumes a housing lifetime of ten years, and thus this level would represent thirty man-years of effort to provide for original construction, maintenance, and utilities. Since houses customarily provide for two or more persons, this seems like a plausible amount.

For clothing, 2065 kilocalories per day is allotted. This amount seems generous, at least in theory. It could be thought of as one person working for three months to produce the raw materials and another working three months each year just to provide a wardrobe for a single person. Our present clothing energy expenditures do not exceed this amount.

Hafner allows 4130 kilocalories per day for transportation. At present transport efficiencies, this would mean an average of 14 miles per day by bus or 17 miles per day by train (or even 92 miles per day by bicycle), but the one-man-one-car transportation system could not be accommodated. At the higher efficiencies that are theoretically possible, the mileage allotment would rise to 50 miles per day by public transport. Clearly, Hafner's estimates would work only if we could revise our present practices in city design.

Beyond the basic necessities are many things: education, recreation, the arts, and the many other ways of seeking satisfaction and meaning in life. For these leisure activities, Hafner allows as much as for food and shelter combined—12,400 kilocalories per day.

To summarize Hafner's minimum energy budget for a satisfactory life:

Food	6,200 kcal/day
Housing	6,200 kcal/day
Clothing	2,065 kcal/day
Transportation	4,130 kcal/day
Leisure	12,400 kcal/day
Total	30,995 kcal/person/day

This total of about 31,000 kilocalories per day is not the only such budget that could be devised. By contrast, present per capita energy use in the United States is about six times as large as in the Hafner minimum. Clearly, more work is needed in this area to find out, even theoretically, what is required for a livable society.

At what levels of energy consumption is life tolerable for everyone? When that question is answered, we must redirect our pattern of energy use so that the desirable level of energy consumption can be reached by everyone. There are extensive political disputes about how much we should assist the poor countries in reaching a higher standard of living. If the rich industrialized countries adopt policies for growth and energy use which foreclose the supply of energy, then arguments about assistance for the poor of the world are just ritualized words. So, on at least two accounts we need to take the above question seriously: first, because our aspirations as human beings and as a nation depend on the possibility of an opportunity for a better world; and second, because a world in which a majority of the population has no chance of achieving a decent standard of living is, almost without question, a world ripe for intolerable conflict.

Figure 12-4 summarizes some of the possibilities. Future population levels used in calculating figure 12-4 represent the medium projection of the United Nations. It is interesting that actual world energy consumption crossed the Hafner minimum in the early 1960s. The future increases in Hafner's minimum were meant to reflect only future population increase. Thus the actual world increases indicate the expanding average per capita usage more than population growth (just as is the case in the United States, where three-fourths of the growth in energy use represents per capita increases). In many areas of the world the expansion in per capita energy use reflects the improved conditions under which many more people live. Thus we should not be surprised to find energy use increasing faster than population in poor nations, for it accompanies any improvement in their standard of living.

But the matter is quite different for the industrialized nations. As we tried to show in chapter 9, much of the increase in per capita energy consumption reflects an elaboration of standard of living for people already well off even by United States standards. Some of this increase, such as the continuing enlargement of auto engines, appears absolutely pointless. The endless expansion of per capita energy consumption by the well-to-do must stop. This need not result in an end to social improvement (as some have suggested) or repression of individual choice. A program to inhibit growth in per capita usage by those with an adequate standard of living could include some or all of the following measures: 1) Increased energy prices to reflect the true costs of energy production—including the environmental costs. 2) An end to tax subsidies to the energy industry. This would be reflected in increased energy prices, and thereby the consumer would be offered a clearer choice of how much energy he wishes to purchase. 3) The choice of rate structures for electricity, gas, and possibly other fuels to provide increased costs for large users whether residential, commercial, or industrial. In principle, rapidly escalating rate structures for large users can be chosen to obtain any decrease in usage desires. 4) Tax incentives or government assistance programs to encourage expenditures for energy conservation (such as better insulation, fewer energy-intensive industrial practices, and the like). 5) Direct regulation or control of expanding energy uses which are unresponsive to other measures (for example, the growing size of automobiles). 6) Promotion of low-energy options for transportation and a halt to subsidies for energy-consumptive transportation. This would require mass transit and efforts at urban redesign. Urban rebuilding will take time, but the present growth rate of urban populations suggests our urban areas will nearly double in the next thirty years, providing an opportunity to improve many items. 7) Tax incentives or direct subsidies to encourage the installation of non-fuel energy systems (such as solar space and hot water heating, windmills, garbage burners, methane generators from garbage, and so on).

Some of the above measures do place restrictions on people, but we must learn, sooner or later, that restriction accompanies a technologically complex society. There is ample individual choice remaining in the above measures. Only the first two items are regressive. If they interfere with social gains, the situation can be remedied with tax or subsidy measures. Maintaining false prices for energy is not the best road to social improvement. Many European countries have done at least as well as the United States in promoting a better standard of living for the poor, although European energy prices have usually been higher than ours.

The question of how much energy the world needs thus cannot be answered directly. Figure 12-4 shows some possibilities for the years 2000 and 2050, based upon the median estimates for world population on those dates. While the undeveloped nations of the world struggle with their many problems, the advanced nations can and should strive to inhibit further growth in per capita energy use. There seems to be no technical

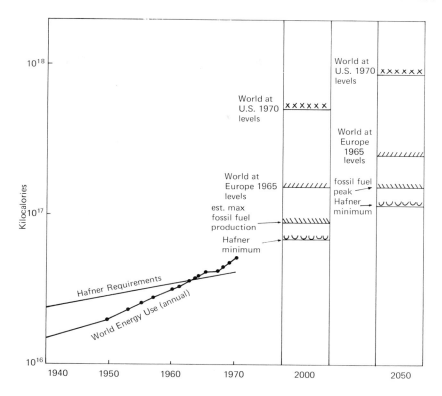

Figure 12-4. Comparison of actual world energy use and requirements of the Hafner minimum. Projections for the years 2000 and 2050 were calculated on the basis of medium United Nations population estimates. World requirements at the per capita use levels of Europe, 1965 and the United States, 1970 were calculated on the basis of the same population projections. Estimated maximum fossil fuel levels are shown for comparison. Sources: S. H. Schurr and B. C. Netschert, *Energy in the American Economy, 1850-1975* (Baltimore: Johns Hopkins Press, 1960); U.S. Department of the Interior, *United States Energy Through the Year 2000* (Washington, D.C.: 1972); J. Darmstadter et al. *Energy in the World Economy* (Baltimore: Johns Hopkins Press, 1971).

reason why energy cannot be supplied at any of the levels shown.

The larger estimates of energy needs in figure 12-4 imply a middle class world on the average. It does not necessarily mean a drably uniform distribution. With an average energy use at the level of 1965 Europe, considerable diversity would be possible, with the lowest energy use levels still offering an adequate standard of living.

All of this should not be taken to suggest that we can achieve these levels in fifty or even one hundred years. The poor nations face staggering problems, of which energy supply is far from the most pressing. Glenn Seaborg, Nobel laureate and former chairman of the Atomic Energy Commission, gives a good example of the proportions of their difficulties by noting that just in the next decade they face the problem of providing new jobs for 300,000 people each week. To do this he suggests their best strategy may be more labor-intensive and less capital-intensive (and energy-intensive) industrial development, including more labor-intensive agriculture.[3] Labor-intensive industries and agriculture can be just as productive as the more energy-intensive kinds (see chapter 4). But if the advanced nations continue blindly to increase their energy consumption, they will eventually foreclose the chances of the poor nations.

The ultimate limits to the admittedly finite capacities of the earth have been a subject of fascination and dispute at least since the time of Malthus, 150 years ago. We know well enough that the limits are there, but the attempts to specify the earliest limits the world might encounter have been inconclusive. It is easy to specify some limits that cannot be exceeded. For example, when the production and transport of fuel to its final use requires more energy than is originally available in the fuel, we will have reached one of these limits. Similarly, when the flux of man's energy use approaches the flux of solar energy driving the physical phenomena of the planet, we will be in serious difficulty. Energy use at this level in a few centuries could be predicted on the basis of simple extrapolations of the growth of the past century and a half. But it seems quite clear that we will not continue to that particular fate (such an extrapolation would include a world population more than one hundred times as large as the present one). Figure 12-3 presents this view of limits for the fossil fuels. It summons up the vision of the cumulative energy use of the world colliding with the fossil fuel limits. But long before that could occur, costs, difficulties of extraction, pollution loads, and other problems will cause a lessening in the use of fossil fuels.

We must conclude that, for the present, the dispute about what limitations we will first encounter for total energy use is not resolvable. In earlier chapters we have tried to display the extent of this dispute as it relates to climate modification, pollution loading, and other possible limits. Limits do exist, and any prudent person would do well to anticipate the worst—and be pleasantly surprised if the larger limits turn out to be correct. The distant future, about which we can estimate little, will probably be better off if we adopt that strategy. It seems unlikely, on the basis of past experience, that we will encounter ultimate limits in a great cataclysm.

The argument about limits to energy use (and other ultimate limits to growth, for that matter) should be watched closely. For anyone outside the narrow circle of anointed experts, the technical issues are difficult or impossible to unravel—depending, as they often do, upon fine technical points or complicated models. Nevertheless, there are several common forms of argument against which anyone can be on guard. Those who argue that there are no real limits often erect a "straw man" limit and then knock it down to prove that there are no limits. One expert, for example, points out that at one hundred times the present level of energy consumption, we would only be using, and releasing, 0.4 percent of the solar energy budget.[4] But this is more energy than drives all the winds and circulation of the atmosphere, and it would be released in highly concentrated areas—not uniformly over the surface of the earth. We cannot say whether this level of energy use would be intolerable, but the argument above does not prove anything at all. Similar overstatements and unsupportable arguments are offered by the prophets of impending doom—with similar assurance. Until these arguments about limits can be resolved (which may take a very long time), we are forced to plan a strategy in the absence of complete information. In such cases, the only sane strategy is a conservative one. The penalty for overestimating the supplies is so great that erring by underestimating the size of the energy supply is a safe, if inconvenient, alternative to disaster. As long as innovation continues with this strategy in mind, there is no reason why social goals need be limited by the availability of energy. A world operating at average 1965 European levels (or even 1970 United States levels) of energy use should be more than comfortable. There is no technical reason why such a strategy cannot proceed. If the world's institutions cannot adapt to these conditions, we are probably undone.

What are our options for long range energy supply? In chapters 5, 6, and 7 we outlined in some detail the present status and future prospects for the main energy sources and those matters will not be repeated here. But it is appropriate to make a summary comparison of our future technical options and estimate how those options might serve us in a future more distant than the ten to fifty year periods for which estimates are most often made. As indicated above, precise estimates of future levels of use or need are not yet possible, but we do know that world energy consumption will continue to increase for some time and that decisions made in the next few decades may have consequences that extend for several thousand years into the future. In these cases, we must make the best evaluation we can despite the uncertainties.

One message from figure 12-3 is that the world will change its main source of fuel or its growth pattern or both. From the point of view of energy supply, there are several options. Because of the fundamentality of energy as a primary input to modern society, the choice among these options is vastly more than a simple engineering or technical choice. Upon this choice, and several others, may depend the very texture of future society. Nevertheless, we must postpone such speculation for yet another moment to summarize some of what has been said along the way about the technical alternatives.

As we have tried to demonstrate, large-scale dependence on liquid petroleum and natural gas is even now, before the peak of world production, coming to a close. It will not do so with the clanging discontinuity implied by simple extrapolation of the cumulative use lines on figure 12-3 — at least not on technical or economic grounds, for the enterprises that manufacture a host of materials from petroleum and natural gas can and will pay higher prices. Even among fuel users, those not easily shifted away from petroleum and natural gas will have to pay higher prices. Cumulative world and United States energy use is likely to pass beyond the limits on the right of figure 12-3 not by exhausing them in succession, but by diverting the remaining supply into the uses which can bear the highest prices.

But the situation for coal is somewhat different. We are still so far from the coal limits—or, more accurately, so early in the total production curve—that we could shift the world to a coal-based economy and continue past growth rates for a few decades, or perhaps for a few centuries, before encountering stringent limits based on availability alone. Table 12-1, which summarizes the future options briefly, shows several reasons why this is not an attractive idea. First, the pollutants associated with coal are troublesome. This pollution is not beyond hope of technological control, but this effort itself will require expenditure of still more energy, and always with the guaranteed failure of achieving complete control. Second, the distribution of coal in the world is not suitable for the present distribution of political control, and the possible conflict over control of energy supplies encourages the same strains that have riddled the past with endless wars. Third, the supply of coal represents the raw material, as a substitute for oil and gas, for many kinds of manufactured chemicals and other goods. We may very well wish to expand (or we may have to expand) our use of coal during the transition to a new mix of energy supplies, but it should be with the understanding that it is only a temporary measure.

For many years the United States has set its hopes for long-range energy supply on the breeder reactor. Problems have arisen and been solved, only to have others arise. Meanwhile the early estimates of cost have doubled and then doubled again. It should be stated here that there are no technical problems that are clearly beyond solution, but there are problems about which there is vigorous dispute. These disputes—between men of considerable understanding of what they are talking about—usually revolve about whether solutions will be too costly in money or dislocation of society or both.

Starting with the observation that the present generation of burner reactors cannot be sustained in the long run without the unimaginable destruction inherent in pursuing ever lower grade uranium ores (and perhaps not even then), we can summarize the main questions remaining about breeder reactors. First are questions about safety. The penalties for a major accident are great. While no accident rate can ever be reduced to zero, nagging fears about the safety of present and projected plans persist. Second, the storage of nuclear wastes and radioactive materials from abandoned plants will be a problem for thousands of years. We are unsure that we can or should make such a commitment on behalf of future society. Third, the widespread availability of plutonium from breeders endangers society with purposeful damage from the maniacal fringe. Even nations from time to time display a kind of madness. Widespread production of plutonium, in the opinion of some, could plunge us into a nuclear nightmare.

It is too early to say the final word on the breeder because the extensive experience that might resolve some of the uncertainties is not available. It would be foolish to discontinue work on breeder reactors at this time or until there are other assured energy sources. It would be even more foolish to place sole reliance, or even primary reliance, on breeder technology until there is a resolution of the present disputes.

If we place fusion and solar conversion side by side, the extremes of the options open to us can be seen. Although table 12-1 summarizes the facts, several points require emphasis. Perhaps because direct solar conversion and nuclear fusion are often talked of as options for the distant future, the idea is often conveyed that they present equally difficult and possibly unsolvable technical questions. As we explained in chapter 6, the fundamental problems of a sustained, controlled fusion reaction are far from solved. Until there is a workable laboratory scheme, all of the engineering problems cannot be specified in detail. But since the operating conditions for a fusion reaction can be specified, some engineering analysis has been done and the problems appear staggering although not beyond the range of possible solution. Resource problems may arise in connection with the engineering solutions, even if the basic problems of sustained fusion reaction are solved.

The situation for solar conversion is quite different. There are no basic research problems to be solved, nor do complex engineering difficulties remain. The only problem with solar energy conversion is that it is thought to be too expensive. Both basic research and engineering improvements could be of great help in reducing the cost, but even at the present cost it may become economically competitive, depending on how high conventional energy costs (including environmental costs) rise. It is probable that solar energy costs could be reduced simply from mass production savings in unit cost.

As table 12-1 shows, there are two kinds of direct solar energy conversion in question. Local conversion for space heating and hot water heating is done now—more widely in other countries than in the United States. It has been shown to be competitive with residential electric heating at 1970 energy prices in places as far north as Boston. Its competitive position will only improve as energy prices increase,[5] even if there are no cost reduc-

Table 12-1.   Long-range Energy Sources

Environmental Problems

	Supply Available	*World Distribution*	*Pollutants*[1]
Coal	55×10^{19} kcal	Spotty. Some areas with little or none.	SOx, COx, NOx, particulates trace elements
Breeder Reactors	Very large	Widespread (but some areas with very low grade ores)	Radiation release from accidents, transport, or storage
Fusion Reactors (H^2-H^2 type)	Very large[3]	Widespread (requires ocean access)	None; but plant and/or coolant may become radioactive
Local Solar Conversion	Very large	Everywhere	None
Central Solar Conversion	Very large	Everywhere	None

1 All methods except local solar conversion generate thermal pollution. Nuclear reactors have a larger heat disposal problem than coal-fired power plants. The amount from fusion reactors and central solar converters is not certain until the technology is specified, but could be largest of all.

2 Engineering and perhaps some basic research associated with removal of pollutants are required, especially for sulfur oxides. It is not clear whether new research and new methods or better engineering of existing methods are required. These problems apply to oil as well as coal.

tions for solar heating units. Experience suggests that, by putting such conversion units into mass production, price can be reduced between twofold and fivefold.

For central solar conversion the engineering problems associated with energy storage need work, but even without dramatic progress there, we could use the energy to produce hydrogen, which is a satisfactory, and environmentally clean, fuel. Roger Schmidt of Honeywell, appearing before

Environmental Problems

Land Use	Basic Research Required	Engineering Development Needed	Scarce Resource Requirements
Extensive and possibly destructive from mining, power stations and storage	None[2]	Cheaper methods of gasification and liquefaction	None
Extensive for mining, processing, and storage and disposal of waste	None	Several large scale engineering improvements needed. Especially for transport and storage safety.	None
Uncertain, probably less than above	Several unsolved and possibly unsolvable problems	Many difficult engineering problems	Niobium or several other scarce minerals have been suggested by some designers[3]
Minimal	None	Engineering for cost reduction for converters[4]	None
Extensive probably larger than breeders or fusion; but not destructive	None	Engineering for cost reduction for converters and energy storage[4]	None

3 If fusion reactors of the lithium-deuterium type are developed, rather than the deuterium-deuterium type, lithium resources may be sharply limiting for fusion power.

4 Photovoltaic conversion would be more attractive if its cost could be reduced to between 1 and 10 percent of its present cost. Some could come from large-scale industrial production of solar cells, but research and engineering breakthroughs may be needed. New or improved energy storage methods are needed if solar conversion to electricity is undertaken.

the House Science and Astronautics subcommittee, estimated that within ten years costs for solar units will be about one thousand dollars per kilowatt of installed electric capacity. This is about twice the present cost of burner reactors, and the breeders will cost even more. For 1973, estimates of the cost of the prototype breeder reactor were two thousand dollars per kilowatt (about double the early 1972 estimates).[6]

There are two less obvious advantages of direct conversion of solar energy. First, it is a low technology option. The poor nations, which often have limited technical capacities, could use this energy source without dependence on the technology of the advanced countries. Second, since the land area involved in solar conversion (of the central station type) is proportional to the energy wanted, there is a built-in warning system for society when energy use begins to get on toward the misty, but real, ultimate limits. By contrast, a successful fusion reactor, located on the sea bottom (more science fiction than plan at the moment), is out of sight and out of mind. In that case, the past suggests that the first notice of too much energy use will come from nature—possibly in the form of disastrous change.

Finally, fusion energy represents the conditions on the sun. We have no real idea how difficult or undesirable it might be to transfer those conditions to the earth even on a small scale (if you can call the world's energy supply a small scale). But the sun is its own fusion reactor and solar energy rains down upon all of us. If it is a little more work to collect it, or even a little more expensive, at least we know what we are doing. From a functional point of view, we will be taking a lesson from the green plants. We will photosynthesize, although not with chlorophyll.

If we, as authors, have a bias, it is for solar energy. We state it without shyness for the reasons outlined above. The level of work on this energy source is expanding, but should be expanded still faster. Work on fusion should continue as well, but the first priority for direct solar conversion should be made clear. At some time in the distant future, society may find it advantageous to use all of the above sources for special applications, and it would be foolish to discard the options until some insoluble problem is encountered in the technical work. But it is even more foolish to burden the same distant future with endless streams of nuclear waste if we need not, or to wait upon fusion, like Mr. Micawber, confident that "something will turn up."

We have pointed out that future options should be left open, even if there are unsolved technical problems with some of the options. Remembering the value of diversity to survival in natural systems, we should plan our future energy supplies to include some of the smaller sources—wind, falling water, geothermal sources, waste burning, biological sources, and so on—in the places where they are available and appropriate. Because some of these smaller sources are local, they do not require the large-scale complexity of the proposed giant nuclear complexes, for example. If the ever increasing complexity of society is not a mortal danger (as some have warned), neither is it an attractive feature to be sought after. Choosing simple technologies when possible and aiming for simplicity of technical means is not a retreat to the past. Choices for simplicity and maintenance of diversity could and should be the hope for a more satisfying and understandable world of the future.

Change continues, but we should not look to the business world, especially the energy companies, for leadership. With their horizontal and vertical control of the main sources of fuel we cannot expect them to urge new choices. The competitive market of classical economics has long since vanished from the energy field and even interfuel competition is vanishing before the wave of mergers and expansion. The industry seeks to reduce what remaining risks it faces through public subsidy. General Electric suggests subsidies may be needed to make breeder reactors "competitive," and the AEC obligingly puts money in its budget in the event that it becomes necessary to pay such subsidies.[7]

Along the way we have said much about the environmental risks and uncertainties associated with energy use. In a larger sense, there are only two things to do with unwanted byproducts of society's activities: we will either reuse our pollutants or live with them diluted into the environment around us.

For energy and fuel use, there are two classes of pollutants—the various chemical and radioactive materials, and heat. For the first class, technical means can convert air pollution to solid waste or otherwise capture the unwanted residual, but it does not go away. If we cannot reuse it, we must store it or dump it somewhere. The second item, heat, will be produced and released as the inevitable consequence of energy use. Where and under what circumstances it is released will determine whether and when it becomes a problem. Ultimately—if growth in energy use continues—it will become a problem; first on a local or regional basis, and later on a worldwide scale. It should be high on our priority list to learn, as accurately as possible, when and under what circumstances these problems may beset us.

There are other environmental difficulties. Amount of land used for fuel production and energy conversion and distribution is already a problem in some areas. Such conflicts in land use can be expected to worsen as long as the growth in energy use continues. More than land use is at stake. The attractiveness of our surroundings is also involved. Oil refineries, strip mines, power lines, or electric utilities are not attractive neighbors, yet more and more of us will find one of these next door as energy use grows.

Any program that lessens the growth rate in energy use, that saves energy, or makes possible a satisfactory life without escalating energy use will be reflected in the diminution of the above problems.

We cannot now conclude for doomsday or utopia. The future holds uncertainty and change—and neither more nor less can be said at present.

At the end of his life, Immanuel Kant said that he was still plagued by three questions. They were: What may I know? What may I hope? and What ought I to do? At the profound level that Kant meant these questions, we cannot answer, but in a narrower context we can make some response.

We know that the fossil fuels will diminish in quantity and will eventually cease to be used as fuels. With adequate work we can know when. We know that growth in the production of goods will cease at some time. We can know (but do not at present) when we must prepare for that time. We know that world energy use will level off or decline at some time in the future. We can know—more accurately than at present—when and to what extent. We know that there is no escape from our problems through increased efficiency. We know we will change. We can know how to make these changes in ways that will economize on energy use and still maintain a satisfactory life.

We may hope for a livable world without a warning and a threat from every look at the near future. We may hope for levels of energy use that make the opportunity for a decent standard of living for everyone a reality. If these hopes can be realized, we may then hope for our individual, private visions of a better world.

What we ought to do is to talk frankly and unashamedly about our hopes for the future—to anyone who will listen. What we ought to do is to seek simplicity and economy of means in the face of complexity. We ought to learn from the past but not seek to return to it. We ought to hold the aims of our own lives to ourselves and not surrender our dreams to the imperatives of faceless experts. For energy specifically, we ought to seek help from the solar energy falling upon us, but not rule out other options for our future.

We—your authors—were trained as scientists. The finest achievements of science have been marked by simplicity and elegance. It has never been the goal of science to find just any explanation of the phenomena of the world around us. The explanations that have enriched understanding are the simple ones, whose very economy of means exhibit the elegance of which we speak. The world does not now have the option of discarding technology. Innovation must continue—more in institutions, perhaps, than in things—but it need not generate ever more complexity and growth. If the innovation is guided by the search for simplicity and economy of means in solving our problems, we may yet find it possible for individual life to be dignified and satisfying — for everyone.

References

1. "Blueprint for Survival," *The Ecologist* 2(1972): 1-43.
2. E. Hafner, *An Energy Budget*, mimeographed (Hampshire College, 1971).
3. G. T. Seaborg, "Science, Technology and Development: A New World Outlook," *Science* 181 (1973): 13-19.
4. P. Sporn, "Possible Impacts of Environmental Standards on Electric Power Availability and Costs," in *Energy, Economic Growth, and the Environment*, ed. S. Schurr (Baltimore: Johns Hopkins Press, 1972), pp. 69-88.

5. R. A. Tybout and G. O. G. Löf, "Solar House Heating," *Natural Resource Journal* 10 (1970): 268-326.
6. R. Gillette, "One Breeder for the Price of Two?," *Science 182 (1973): 38.*
7. *Ibid.*

FOR FURTHER READING

Borgstrom, G. *The Food and People Dilemma.* North Scituate, Mass.: Duxbury Press, 1973.

Ehrlich, P., and Ehrlich, A. *Population, Resources, Environment.* San Francisco: Freeman, 1970.

Foreman, H., ed. *Nuclear Power and the Public.* Garden City, New York: Anchor Books, Doubleday & Co., Inc., 1972.

Garvey, G. *Energy, Ecology, Economy.* New York: W. W. Norton & Co., Inc., 1972.

Georgescu-Roegen, N. *The Entropy Law and the Economic Process.* Cambridge, Mass.: Harvard University Press, 1971.

Harte, J., and Socolow, R. eds. *Patient Earth.* New York: Holt, Rinehart and Winston, Inc., 1971.

Holdren, J., and Herrera, P. *Energy.* San Francisco: Sierra Club, 1971.

Hottel, H., and Howard, J. *New Energy Technology — Some Facts and Assessments.* Cambridge, Mass.: MIT Press, 1971.

Inadvertent Climate Modification: Report of the Study of Man's Impact on Climate. Cambridge, Mass.: MIT Press, 1971.

Jantsch, E. *Technological Planning and Social Futures.* New York: John Wiley & Sons, 1972.

Jungk, R. *Brighter Than a Thousand Suns — A Personal History of the Atomic Scientists.* New York: Harcourt, Brace & World, Inc., 1958.

Laurence, W. *Men and Atoms — The Discovery, the Uses and the Future of Atomic Energy.* New York: Simon and Schuster, 1962.

Lilienthal, D. *Change, Hope, and the Bomb.* Princeton, N.J.: Princeton University Press, 1963.

Man's Impact on the Global Environment. Report of the Study of Critical Environmental Problems. Cambridge, Mass.: MIT Press, 1970.

Matthews, W. *et al.,* eds. *Man's Impact on the Climate.* Cambridge, Mass.: MIT Press, 1971.

Matthews, W. *et al.,* eds. *Man's Impact on Terrestrial and Oceanic Ecosystems.* Cambridge, Mass.: MIT Press, 1971.

Mishan, E. *The Costs of Economic Growth.* New York: Praeger, 1967.

Mumford, L. *Technics and Civilization.* New York: Harcourt, Brace, & Co., 1934.

National Academy of Sciences-National Research Council. *Resources and Man.* A Study and Recommendations by the Committee on Resources and Man. San Francisco: W. H. Freeman and Co., 1969.

Odum, H. *Environment, Power, and Society.* New York: Wiley-Interscience, A Div. of John Wiley & Sons, Inc., 1971.

Reynolds, J. *Windmills and Watermills.* London: Hugh Evelyn Ltd., 1970.

"The Energy Crisis, Parts I, II and III," *Science and Public Affairs* Bulletin of the Atomic Scientists 27, Nos. 7, 8, and 9 (1971).

Scientific American. 224, No. 3. (September, 1971).

Stacks, J. *Stripping.* Sierra Club, San Francisco. 1972.

Steinhart, C., and Steinhart, J. *Blowout: A Case Study of the Santa Barbara Oil Spill.* North Scituate, Mass.: Duxbury Press, 1972.

Szczelkun, S. *Survival Scrapbook #3: Energy.* New York: Schocken Books, 1974.

United Nations, Proceedings of the United Nations Conference on New Sources of Energy. *Geothermal Energy.* Vol. 2 and 3. New York: United Nations, 1964.

United Nations, Proceedings of the United Nations Conference on New Sources of Energy. *Solar Energy.* Vol. 4, 5, and 6. New York: United Nations, 1964.

United Nations, Proceedings of the United Nations Conference on New Sources of Energy. *Wind Power.* Vol. 7. New York: United Nations, 1964.

United States, Environmental Protection Agency. *Alternative Futures and Environmental Quality.* Washington, D.C.: 1973.

INDEX

of, 27-29, 32; energy loss in producing, 166; and environment, 149-150; geothermal, 133, 135; for home heating, 182-183; from hydroelectric plants, 127, 128, 129, 130; from nuclear reactors, 104, 106, 114, 115; power plant siting for production of, 148-151; from solar power, 143, 144; from tidal power, 126; transmission of, 158-163; thermoelectric generation, 165; use of, 10, 147-148, 171, 172, 173; variations in demand for, 163; water used in production of, 152; from windmills, 123

Electric motor: 11, 28, 32

Electric utility industry: 221

Electrolysis: 124, 144

Electrons: 97, 98

Energy: balance of, and atmospheric components, 203; chronology of man and, 9; classifying use of, 186-190; control of resources of, 219-223; demand for, 28-30, 163; in ecosystems, 39-40; effects of agriculture on balance of, 208, 210; exponential growth in use of, 33; and food system, 38, 39, 42-47; forms of, 23-25; future needs for, 246-252; future use of, 241-246; history of use of, 7-17; long-range sources of, 252-259; man-made and natural, compared, 207; minimum per capita requirements of, 248-249; options for national policy for, 236-239; parties interested in policy for, 234-236; present policy for, 234-236; priorities of controllers of, 223-230; reducing demand for, 57-60, 191-196; required for cooling systems, 155; and society's goals, 218-219;

storage of, 29, 123, 137, 139, 144; use of, broken down by sectors, 171-186; wasted, 26, 27, 28, 151-153

Environmental problems: 81-94, 149-150, 151-155, 229, 259

Environmental Protection Agency: 110, 234

Environmental standards: 229

Equisetum: 74

Erosion: 81, 82, 83

Escherichia Coli: 34

Eutrophication: 93, 154

Evaporation: energy loss through, 153, and heat balance, 210; solar, for producing salts, 137; in spray canal, 157; in wet cooling tower. 156

Evaporation ponds: 137

Farming. *See* Agriculture

Fast breeder reactor: 103, 104

Federal Power Commission: 104, 220

Fermi, Enrico: 98-99, 101

Fertilizers: 45, 58

Fire, discovery of: 20-21, 206

Fishing industry: 56

Fission reactions: 100, 102, 115, 116, 117

Flat plate collectors: 142-143

Flourescence: 97

Flushing time (lakes): 111

Focusing collector: 143-144

Food: as energy, 37; and energy use, 43, 45. 47, 48, 55-61, 172, 186; expense of, 44-45; for an industrial society, 43-52, 187; processing and packaging of, 43, 46, 59, 172, 175; shortages of, 60; solar drying of, 137; transport of, 44, 47, 59; world supplies of, 53-55, 60

Food industry: 42-60, 172, 175,

Volcanoes: 3, 131, 204

War: energy distribution and, 5; and technology of weapons, 16

Waste heat: 30; biological effects of, 153-155; and climate, 206-208, 210; disposal of, 155-157; from electric plants, 151-158; from geothermal electricity, 133-134; from hydropower, 127, 129-130; recycling of, 157-158; uses for, 158

Wastes: disposal of, 48, 90, 120-121, 152, 155-157; garbage, 120-121; gas, 70, 72, 190, 191; oil, 90; radioactive, 113, 114, 115, 255; water, 134

Water: for cooling, 149, 152; and energy balance, 210; heated (hydrothermal), 132; hydrogen derived from, 124, 144; pollution of, 82, 83, 85, 86, 87, 88, 89, 90, 130, 145; thermal pollution of, 129-130, 151-155

Waterfowl: 91

Water freight: 181

Water heating: 29, 137-138, 140, 183, 186, 192, 244

Water power: 17, 127-131, 150 163

Water vapor: 203

Waterwheels: 14, 15, 127

Watt, James: 27

Weapons, nuclear: 95, 99, 100, 107, technology of, 16

Wet cooling tower: 155, 156

Whale oil: 66

Windmills: 15, 122-125

Wind power: 17-18, 59, 121-126

Wood, as fuel: 120

Work: heat and, 22-24

X rays: 97

Zinn, Walter: 99